MALCOLM PIRNIE, INC.
LIBRARY COPY
COLUMBUS, OHIO

PROCESS MODIFICATIONS FOR INDUSTRIAL POLLUTION SOURCE REDUCTION

By Lawrence L. Tavlarides, Ph.D.
Pritzker Department of Environmental Engineering
Illinois Institute of Technology
Chicago, Illinois

INDUSTRIAL WASTE MANAGEMENT SERIES
James W. Patterson, Executive Editor

LEWIS PUBLISHERS, INC.
121 SOUTH MAIN STREET □ P.O. DRAWER 519 □ CHELSEA, MICHIGAN 48118

Library of Congress Cataloguing in Publication Data

Tavlarides, Lawrence L., 1942–
 Process Modification for Industrial Pollution Source Reduction.

 (Industrial Waste Management Series   )
 Bibliography: p.
 Includes index.
 1. Factory and trade waste—Environmental aspects. 2. Pollutants—Environmental aspects.
I. Title. II. Series
TD897.T333 1985   604.6   84-21257
ISBN 0-87371-003-7

COPYRIGHT © 1985 by LEWIS PUBLISHERS, INC.
ALL RIGHTS RESERVED

Neither this book nor any part may be reproduced or transmitted in any form or by any means, electronic or mechanical, including photocopying, microfilming, and recording, or by any information storage and retrieval system, without permission in writing from the publisher.

LEWIS PUBLISHERS, INC.
121 South Main Street, Chelsea, Michigan 48118

PRINTED IN THE UNITED STATES OF AMERICA

PREFACE

This volume, <u>Process Modifications for Industrial Pollution Source Reduction</u>, describes one of the most innovative and potentially advantageous modern concepts for industrial waste management. It was prepared particularly for chemical and other engineers who work in industry and seek direction in process modification approaches for pollution abatement. The volume will also be extremely useful to environmental engineers, who increasingly must deal with pollution generation at the manufacturing point source. This volume will also assist the students in understanding the linkage between manufacturing technology and industrial pollution, and demonstrate how process modifications can effectively reduce or eliminate pollutants.

Process modification can be quite site specific, depending upon the category of industry, particular raw materials and products, and the mode of manufacture. To accomplish this study, Dr. Tavlarides has selected a spectrum of types of industry and products, dissected each manufacturing step, and identified numerous process modifications. In many instances the modification can be accomplished with simple, proven technology, and with metallurgical and chemical engineers, professors and their significant cost advantage over conventional pollution control methods. Where modifications are less well established, directions for research, development and demonstration have been suggested. This volume convincingly demonstrates that in many industries process modification is a technically feasible and economically attractive approach to industrial waste management.

Review of this work by the USEPA does not necessarily indicate approval of its contents. Special appreciation is expressed to Mr. William A. Cawley, Deputy Director, USEPA Industrial Laboratory, and Chairman of the IWERC Policy Board.

James W. Patterson, Ph.D.
Series Executive Editor and
Director of
The Industrial Waste Elimination
Research Center

SERIES PREFACE

The philosophy of industrial waste management has changed rapidly in recent years, due to regulatory and economic incentives resulting from increasing restrictions on air, water and hazardous wastes pollution, and from escalating costs of pollution abatement. Traditional methods of industrial pollution control often ignore alternative management strategies with significant economic and environmental benefits. Specifically, opportunities for source reduction, by-product recovery, and recycle and recovery for reuse have historically received scant attention. Today, industry is increasingly considering such alternative approaches, and finding many opportunities for their cost effective implementation.

The Industrial Waste Elimination Research Center (IWERC) is a consortium of Illinois Institute of Technology and the University of Notre Dame, established under the auspices of the U.S. Environmental Protection Agency (USEPA), Office of Exploratory Research. It operates under the supervision of a Policy Board, and with advice from its Scientific Advisory Committee and Industry Advisory Council. IWERC has the mission to identify management strategies, and to perform technology evaluation, research, and development on innovative alternative technologies for industrial waste management. As part of its program, the Center commissioned comprehensive studies in three areas:

1. Recovery, recycle and reuse of industrial pollutants;

2. Management of industrial pollutants by anaerobic processes; and

3. Process modifications for pollution source reduction in chemical processing industries.

These studies have yielded state-of-the-art information on opportunities for applications of alternative strategies for industrial waste management. In view of the current widespread and growing interest in such alternative approaches, the results of these studies are being made available, as part of this Industrial Waste Management Series.

                              James W. Patterson, Ph.D.

LIST OF TABLES

Table		Page
1.1	Flow Rates of Metals in Process Streams	9
1.2	Concentration Ranges in Effluent Streams From Electrorefining Process	10
1.3	Pollutants and Suggested Control Strategy for Copper Production by Pyrometallurgical Processes	11
1.4	Pollutants and Suggested Control Strategy for Copper Production by Hydrometallurgical Processes	14
2.1	Composition of Raw Waste Streams From Common Metals Plating	27
3.1	Approximate Trace Element Analysis of Coal Pretreatment Dust After Wet Scrubbing	34
3.2	Estimated Trace Elements Composition of the SRC Liquefaction Residue	35
3.3	Pollutants and Control Options in SRC Liquefaction Process	38
3.4	Laboratory Leaching Results of Chem-Fixed Refinery Wastes	42
3.5	Spent Harshaw Nickel Catalyst Analysis	43
3.6	Distribution Coefficients for Various Phenols in Butyl Acetate at 300°K	44
3.7	Control Options for the Concentration Acid Gas Stream	45
3.8	Pollutants and Control Options in Lurgi SNG Process	46

Table		Page
4.1	Pollutants and Control Options in Nitric Acid Production Process	57
4.2	Pollutants and Control Options in TNT Production	62
4.3	Pollutants and Control Options in NC Production	67
5.1	Average Analysis of Quench Water Samples	76
5.2	Coke Oven Gas	79
5.3	Analysis of Blast Furnace Scrubber Wastewaters Flow Rate	80
5.4	Wastewater Thickener Underflow	81
5.5	Raw Wastewaters From Steel Making Operations	88
5.6	Pollutants and Suggested Control Strategy, Iron and Steel Industry	91
6.1	Typical Emissions Rates From Batch Digester In kg Sulfur/Ton of ADP*	102
6.2	Emission Rates From Vacuum Washing In kg Sulfur/Ton ADP	103
6.3	Emission Rates From MEE In kg Sulfur/Ton ADP	104
6.4	Emission Rates From Recovery Furnace	106
6.5	Particulate Emission Rates From the Recovery Furnace	106
6.6	Emission Rates From Smelt Dissolve Tank	109
6.7	Lime Kiln Emission Rates	111
6.8	Pollutants and Control Options in Kraft Pulping Process	112
7.1	Chemical Analysis of Red Muds	119

Table		Page
7.2	Emissions From Solderberg Cell	124
7.3	Effect of Cell Operating Parameters as Flouride Effluent	126
7.4	Suggested Process Modifications for Pollution Control in the Primary Aluminium Industry	128
8.1	Pollutants and Suggested Strategy for Selected Phosphate Rock Fertilizer	144

LIST OF FIGURES

Figure		Page
1.1	Flow Chart Showing the Principal Process Steps For Extracting Copper From Sulphide Ores	4
1.2	Comminution and Froth Flotation	6
1.3	Flow Chart For Electrorefining Process	8
1.4	Flow Chart For Solvent Extraction Process	15
1.5	Process Schematic For Uranium Recovery	17
2.1	Flow Chart For Water Flow in Chromium Plating Zinc Die Castings, Decorative	26
3.1	Flow Sheet For An Integrated Liquefaction Process	33
3.2	Lurgi SNG Process	40
4.1	Processes In The Explosive Industry.	52
4.2	Flow Chart For Nitric Acid Production	53
4.3	Flow Chart For TNT Production	59
4.4	Flow Chart For Nitrocellulos Production	64
5.1	Overview of Iron and Steel Manufacturing Process	72
5.2	Coke Making	74
5.3	Coke By-Products Recovery	78

Figure		Page
6.1	Flow Chart of Kraft Pulping Process	100
7.1	Bauxite Processing	118
7.2	Primary Aluminium Production	122
8.1	Wet Process Phosphoric Acid Production	135
8.2	Flow Chart of Normal Superphosphate Production Process	139
8.3	Flow Diagram For TVA Ammonium Phosphate Process	142

CONTENTS

1. **REFINING OF NONFERROUS METALS** 3

   Introduction 3
   Copper Production By Pyrometallurgical
   Processes 3
   Copper Production By Hydrometallurgical
   Processes 11
   Uranium Production By Hydrometallurgical
   Processes 16

2. **ELECTROPLATING** 25

   Electroplating of Common Metals 25

3. **COAL CONVERSION PROCESSES** 31

   Introduction 31
   Liquefaction 31
   Gasification 39

4. **EXPLOSIVES INDUSTRY** 51

   Introduction 51
   Nitric Acid Production 51
   TNT Production 56
   Nitrocellulose Production 61

5. **IRON AND STEEL INDUSTRY** 71

   Introduction 71
   Coke Making 73
   Coke By-Product Recovery 77
   Iron Making 79
   Steel Making 85
   Acid Pickling 89

6. PAPER AND PULP INDUSTRY 99

Introduction 99
Kraft Pulping 99
Recovery Furnace System 105
Lime Kiln 110

7. THE PRIMARY ALUMINIUM INDUSTRY 117

Introduction 117
Bauxite Processing 117
Primary Aluminium Smelting 121

8. PHOSPHATE FERTILIZER INDUSTRY 133

Introduction 133
Wet Process Phosphoric Acid Production . . 138
Ammonium Phosphate Production 140

INTRODUCTION

The objective of this project on process modifications for pollution source reduction in the chemical processing industries is to identify potential changes in manufacturing processes which could have significant impact on pollution elimination or reduction at its source. Toward this end, eight industries were evaluated to develop a matrix of significant pollution problems and attendant process modifications which would have impact on the reduction or elimination of pollutants inherent in these processes. The industries evaluated are as follows:

1. Refining of Nonferrous Metals
2. The Electroplating Industry
3. Coal Conversion Processes
4. Specialty Chemicals
5. The Iron and Steel Industry
6. The Paper and Pulp Industry
7. The Primary Aluminum Industry
8. Phosphate Fertilizer Industry

Although these industries are diverse, it has become apparent through this study that generic pollution problems cut across most of these and other industries not covered in the study. Accordingly, various process modification strategies and attendant research programs which would minimize these pollution problems and are generic in nature, are identified.

CHAPTER 1

REFINING OF NONFERROUS METALS

INTRODUCTION

This chapter is concerned with the refining of nonferrous metals from mineral ore. Copper and uranium production technology is used to demonstrate the industrial category of nonferrous metal refining. The chapter is divided as follows:

1. Copper Production by Pyrometallurgical Processes
2. Copper Production by Hydrometallurgical Processes
3. Uranium Production by Hydrometallurgical Processes.

COPPER PRODUCTION BY PYROMETALLURGICAL PROCESSES

Ninety percent of the world's primary copper originates in the sulfide ore form. A vast majority of this copper is extracted by pyrometallurgical techniques, because sulfide ores are not easily leached. Typical copper deposits contain only 1 to 2 percent copper, and therefore, the ore must undergo several process steps to concentrate the copper before smelting. This method of copper extraction includes the following steps [Biswas and Davenport, 1980]:

1. Concentration by froth flotation
2. Roasting
3. Matte smelting
4. Converting to blister copper
5. Electrorefining

The pollutants of primary concern include SO_2, particulates, slag, and dissolved metal salts in various effluent streams [Barbour, 1980]. Figure 1.1 is a flow chart of this process, indicating the numerous ways in which copper concentrate can be smelted to produce blister copper.

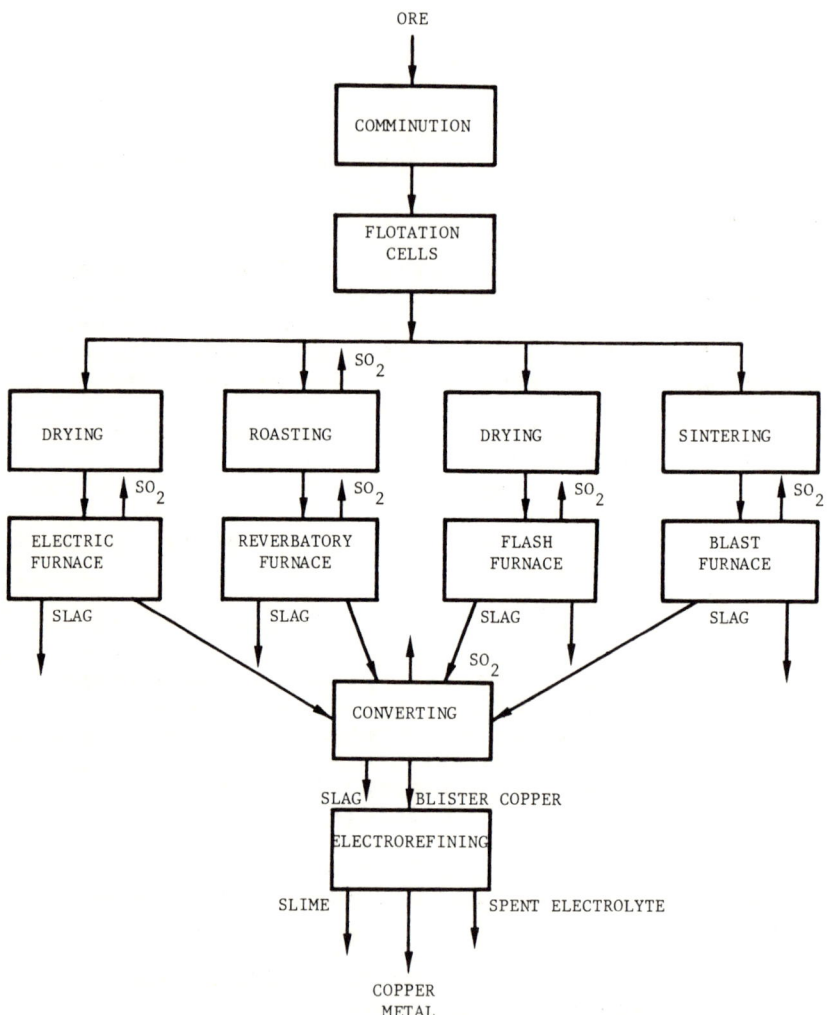

Figure 1.1. Flow Chart Showing the Principal Process Steps For Extracting Copper from Sulphide Ores. (Biswas and Davenport, 1980).

Other sources of wastewater include acid plant blowdown, slag granulation, and cooling of the hot metal anodes. Electrorefining also generates waste through spent electrolyte, washing cathodes, and the recovery of slime [U.S. EPA, 1975].

Comminution and Froth Flotation

A copper concentrate is produced by a series of flotation cells which selectively float copper sulfide to the surface of the bath, so the concentrate can be skimmed. This process is depicted in Figure 1.2. The tailings taken from the tank are sent to evaporation ponds, where the water is clarified and recycled [Biswas and Davenport, 1980].

The tailings require pH adjustments and extraction of arsenic, cyanide, and some metals. The amount of water consumed for this process step can be reduced if improvements are made to the grinding efficiency of the dry ore.

Roasting of Copper Concentrate

The copper concentrate can be refined by several different methods of roasting and/or smelting, as shown in Figure 1.1. Roasting of sulfide concentrates produce calcines for leaching, or for reverbatory (or electric) furnace smelting. The fluid-bed roaster is the best device for sulfide roasting, since it has a high production rate and produces 5-15 percent by volume SO_2 in the flue gases, which can easily be converted into usable forms of sulfur [Gill, 1980].

Smelting

The process of smelting and converting varies widely in equipment design, but the pollutants are essentially the same: SO_2, particulates, and slag. There are several reasons for cleaning gases from metallurgical processing; the most important of these are [Gill, 1980]:

1. To recover valuable particulate material which can be returned to the plant for reprocessing.

2. Environmental pollution control, from the viewpoint of employees exposed to harmful gases and, SO_2 combining with

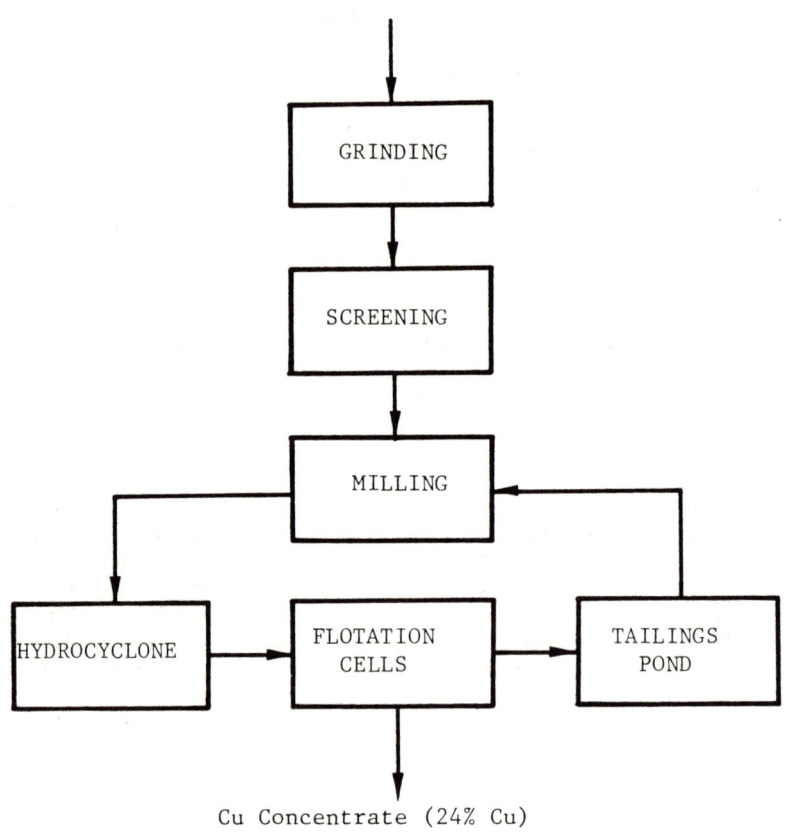

Cu Concentrate (24% Cu)

Figure 1.2. Comminution and Froth Flotation (Biswas and Davenport, 1980).

atmospheric water vapor to form acid rain, damaging plants and injuring animals.

3. To recover valuable gaseous by-products such as SO_2, which can be used as the feed material for the production of marketable sulfuric acid or elemental sulfur.

4. To recover fuel value of gaseous by-products (e.g. combustible CO.)

5. To reclaim the sensible heat to the flue gases in waste heat boilers.

Optimization of the smelter design can reduce the quantity of environmental pollutants, or alter the effluent streams so that the pollutants can be more easily removed. If the SO_2 concentration is greater than 4 percent by volume, it can be converted into sufluric acid without difficulty. Typically, SO_2 is converted into sulfuric acid, sulfur, or ammonium sulfate.

A survey of metals in various process streams was performed at one smelting plant. Table 1.1 is a compilation of flow rates [Schwitzebel, et al. 1978]. Some of the streams were not measured or recorded, but the Table does illustrate the number of different elements present in the effluent streams.

To recover the metals from the captured particulates, it may be possible to use acid leaching, and solvent extraction to selectively remove metals from low concentration leach solutions.

Electrorefining

Electrorefining is the final step of purifying blister copper to 99.99 percent copper, which can be used for most industrial applications. The spent electrolyte is recycled, but some of the acid must be treated to remove excess nickel and unclaimed copper, as well as trace metals. Slime develops from the particulates and precipitates, and can be treated to recover copper and/or precious metals. Figure 1.3 is a flow chart illustrating the representative of the electrorefining process [U.S. EPA, 1975].

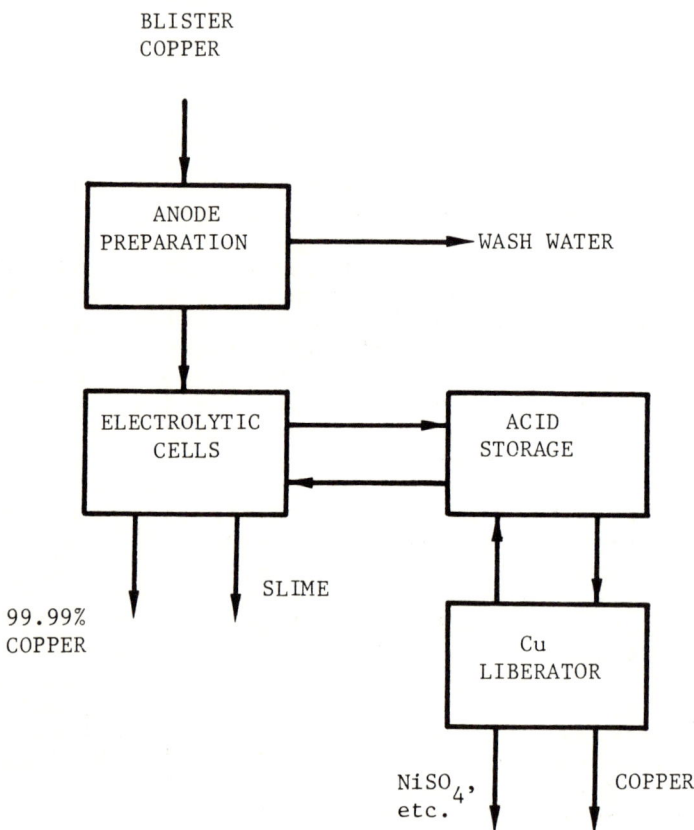

Figure 1.3. Flow Chart for Electrorefining Process (U.S. EPA, Feb. 1975).

TABLE 1.1 FLOW RATES OF METALS IN PROCESS STREAMS

Element	Feed	Matte	Slag	Flue Outlet	ESP Catch
Al	400	17	700	0.10	1.1
As	190	48	48	76	30
Ba	31	33	43	0.64	0.023
Be	0.072	$4 \times 1 \times 10^{-3}$	0.032	0.0034	0.0004
Ca	770	12	1900	0.011	2.1
Cd	59	37	0.33	0.076	0.74
Cr	0.076	0.64	3.8	0.044	0.018
Cu	16000	18000	220	1.8	62
F	3.4	0.012	2.2	9.4	0.032
Fe	10000	11000	12000	0.55	42.6
Hg	0.018	0.020	0.0091	0.033	0.00015
Mo	79	8.5	89	0.17	6.6
Ni	0.70	2.0	0.76	0.011	0.059
Pb	49	84	13	0.38	5.3
Sb	6.0	4.6	3.4	0.030	1.3
Se	10	0.17	5.5	0.65	0.16
Si	1100	42	4800	1.7	1.7
V	0.92	0.33	0.86	0.027	0.011
Zn	42	31	30	0.22	4.6

ESP = electrostatic precipitation

Table 1.2 lists the concentration ranges of several priority pollutants contained in the designated process streams. Slime is either discarded, or undergoes further treatment for recovery of metals. A bleed from the electrolyte, water used to wash the anodes during preparation, and slime are the three primary sources of water contamination from the electrorefining process.

Table 1.3 summarizes the pollutants and suggested control strategies for copper production by pyrometallurgical processes.

Recommended Areas for Process Modification

After evaluating the production technology, the following areas are recommended for further research and development:

Table 1.2 Concentration Ranges in Effluent Streams from Electrorefining Process [U.S. EPA, 1975]

POLLUTANT	ELECTROLYTE (g/L)	SLIME (%)
Antimony	0.2 to 0.7	0.5 to 5.0
Arsenic	0.5 to 12.0	0.5 to 4.0
Copper	45.0 to 50.0	20.0 to 40.0
Lead	–	2.0 to 15.0
Nickel	2.0 to 20.0	0.1 to 2.0
Selenium	–	1.0 to 20.0
Silver	–	3.4 to 27.4

- Optimization of the communition of dry mineral ore, to reduce the amount of water used in froth flotation.

- Recover valuable products by dissolving the metals from particulates collected in the electrostatic precipitators, and use solvent extraction procedures to extract the desired products.

- Use solvent extraction to recover copper or other metals from effluent streams, before the water is sent to the tailing ponds.

- Improve smelter design to contain and eliminate the release of SO_2 and other hazardous flue gases into the atmosphere.

COPPER PRODUCTION BY HYDROMETALLURGICAL PROCESSES

Hydrometallurgy has three advantages over the smelting and converting processes that are traditionally used for the manufacture of copper. First, leaching is a more economical method of extraction when the ore has a low copper concentration. Second, increasing energy costs of the furnaces is making pyrometallurgy less economical. Third, it is becoming increasingly difficult for smelters to meet stringent environmental regulations [Barbour, 1980].

Table 1.3. Pollutants and Suggested Control Strategy for Copper Production by Pyrometallurgical Processes

Process	Pollutant	Source in Process	Nature of Pollutant	Pollutant Control Strategy
Ore grinding	Particulates	Crushers, grinders, ball mill etc.	Inorganic	Electrostatic precipitator
Froth flotation	Metals, Dissolved sludge, cyanide, pH etc.	Tailings	Inorganic and organic additives	Solvent extraction, evaporation, tailings pond
Smelting (a) Reverb (b) Electric (c) Flash (d) Blast	SO_2, slag and particulates	Flue gases, slag dump	Inorganic and Gaseous SO_2	Acid plant for SO_2, ESP, solvent extraction
Converting	SO_2, slag and particulates	Flue gases, slag dump	Inorganic and gaseous SO_2	Acid plant, ESP, solvent extraction
Anode preparation	Dissolved metals, H_2SO_4	Anode wash	Inorganic	Solvent Extraction
Electro-refining	Dissolved metal, H_2SO_4	Electrolyte, cathode cooling etc.	Inorganic	Solvent Extraction

The hydrometallurgical process entails three major steps; leaching, cementation (or alternately, solvent extraction), and electrowinning. This method of extracting copper from mineral ore has not been extensively applied to sulfide ores, which are not readily leached. Due to the environmental problems associated with SO_2, hydrometallurgy is becoming a more acceptable method of producing copper [Biswas and Danvenport, 1980].

Leaching

There are numerous methods of leaching, and several types of leaching solutions. Sulfuric acid is commonly used as a leaching medium, but new technology has introduced other forms of leaching solutions such as ammoniacal and ferric chloride solutions. Leaching of sulfide ores can be improved by bacterial enhancement as well. The pregnant leach solution can undergo either solvent extraction, ion exchange or cementation to recover the copper, while leaving the remaining trace metals in the leach solution [Wadsworth, 1980]. There are four methods of leaching [Wadsworth, 1979]:

1. In situ leaching
2. Dump and heap leaching
3. Vat leaching
4. Agitated leaching

In situ and dump leaching are long term processes which can leach the ore with acid solution for several years. Due to this long period of exposure, significant quantities of the sulfide minerals will dissolve into the leach solution. Contamination of ground water may occur and could produce serious detrimental effects to the surrounding environment.

Vat leaching and agitated leaching are performed with much smaller volumes of crushed ore, at much shorter periods of exposure. These short term processes are useful for ores containing ocpper in the oxide form, which dissolve easily in acid solutions. The pregnant leach solution is taken to refining, while the remaining sludge is either disposed or recycled for additional processing to recover other metals not initially extracted.

If the leach solution has a high concentration of copper (40 to 60 g/l), it can be sent directly to the electrowinning circuit [Flett, 1974]. When the solution has a low concentration, making direct electrowinning uneconomical, then a process step must be taken to precipitate the copper from the rest of the leach solution, and redissolve the copper at a higher concentration into an electrolyte. Three commercially available processes that accomplish this task are cementation, solvent extraction, and ion exchange.

Cementation

Cementation is a reaction where copper deposits directly onto the surface of sacrificial scrap iron. The effluents contain very large quantities of iron salts, and are generally sent to the tailings pond for waste treatment.

Solvent Extraction

Solvent extraction is a process having several circulation loops, where copper ions are selectively extracted from a leach solution by a specific solvent. The extracted copper is stripped from the solvent which is then recirculated into the extraction circuit. The flow chart shown in Figure 1.4 illustrates the multiple circuits of the solvent extraction process.

Some of the pollution problems that exist in this process are the build up of trace metals in the leach solution, and entrained solvent in the extraction and stripping sections. Problems associated with entrainment are complicated by the unavoidable degradation of the solvent. This reduces the extraction efficiency of the organic solvent, and some make up must be added to the recycle.

Table 1.4 summarizes the pollutants and suggested control strategies for copper production by the hydrometallurgical process.

Recommended Areas of Process Modification

After evaluating the production technology the following areas are recommended for further research and development:

PROCESS MODIFICATIONS

Table 1.4 Pollutants and Suggested Control Strategy for Copper Production by Hydrometallurgical Processes

Process	Pollutant	Source in Process	Nature of Pollutant	Pollutant Control Strategy
Leaching in situ, heap, vat, agitated	Dissolved metal salts, sulfuric acid organic additives (wetting agents)	Leach solution	Inorganic, metals and acid	Recycle leach solution after solvent extraction of excess metals, watch for seepage of metals into the ground water
Cementation	Dissolved metal salts	Iron ion from scrap iron	Inorganic	Solvent extraction of effluents
Solvent extraction	Dissolved metals, entrained solvents	Excess from process streams, solvent degradation	Inorganic metals and organic solvents	Improve mixer-settlers, study methods for regeneration of solvents
Electro-winning	Dissolved metals, acid, cyanide	Excess from process streams, cathode wash and cooling	Inorganic metals and acid	Solvent extraction of effluents

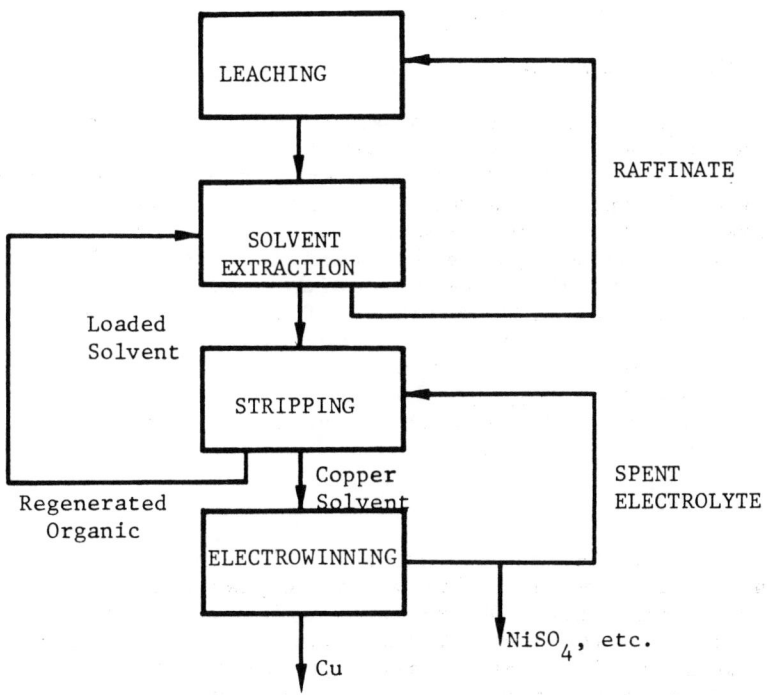

Figure 1.4. Flow Chart of Solvent Extraction Process. (Gill, 1980).

- Recover trace metals in the effluent streams by solvent extraction. Conduct research in the area of simultaneous extraction of several metals.

- Recover entrained solvents by filtration, flotation, centrifugal separation, etc. Evaluate methods of separating liquid-liquid mixtures.

- Evaluate new techniques for leaching sulphide ores or develop new techniques to improve selectivity in leaching mineral ores.

URANIUM PRODUCTION BY HYDROMETALLURGY PROCESSES

Uranium is becoming an increasingly important energy source. Uranium content of a typical ore is in the order of 0.2 to 0.3 percent U_3O_8. For this reason, large quantities of material must be handled to mine and extract uranium. The conventional recovery processes include acid or alkaline leaching, solution purification, and product precipitation into yellow cake (U_3O_8). Solvent extraction and ion exchange have become the primary methods for solution purification and concentration of uranium [Reed, et al. 1979]. Figure 1.5 is a presentation of a process flow chart for the refining of uranium ore.

Leaching

As mentioned in Section 1.3, there are numerous methods of leaching with varying degrees of ore preparation. Interest has increased in the use of in-situ leaching of underground deposits for uranium mining. The advantages of in-situ leaching are the significant reduction of processing costs, and the minimal disturbance to the surface conditions. The subsurface deposit is flooded with leach solution, which is then pumped to the surface for uranium recovery. With in-situ leaching, the ore body remains intact and only the metal is removed. Therefore, a relatively small volume of waste requires disposal.

This method of leaching can only be done when the ore body is contained within a rock formation which is relatively impermeable; otherwise the ground water may become contaminated. The

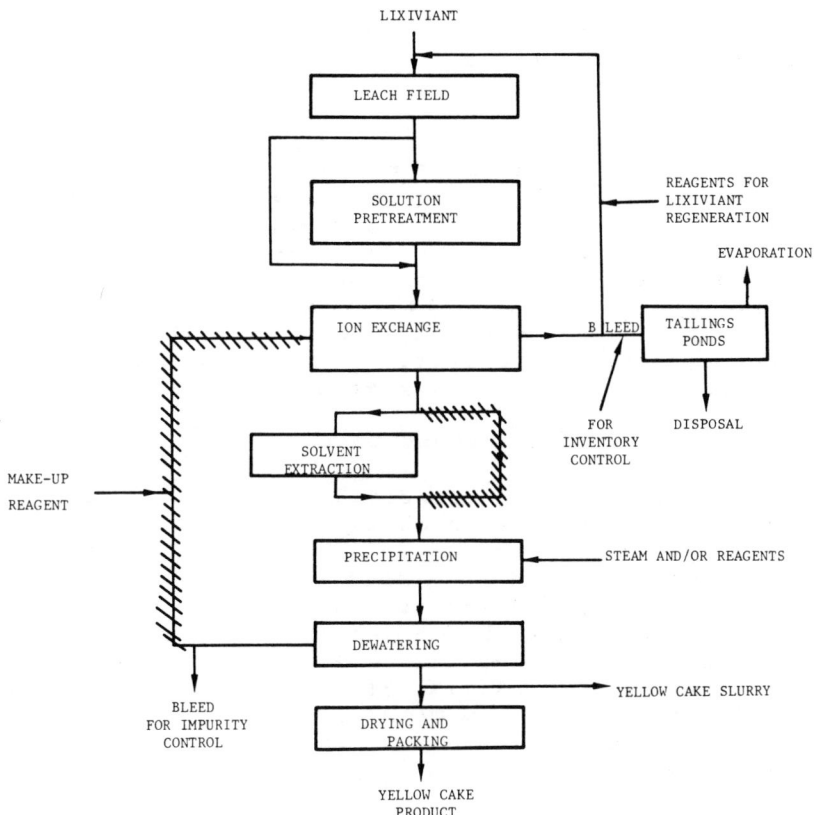

Figure 1.5. Process Schematic for Uranium Recovery

possibility of polluting fresh-water aquifers is present, particularly when a deposit is contained within an aquifer. With proper design, operation, monitoring, and cleanup, in-situ leaching can be an environmentally acceptable method of extracting metals from subsurface deposits.

Other methods of leaching can be used, but the damage to the surface, the amount of waste produced, and the costs for processing the ore increase.

Uranium can also be recovered from leach solutios used for the extraction of other non-ferrous metals like copper, zinc, and nickel. Once the uranium is reclaimed from the leach solution it can be recycled into the leaching process after some "make-up" is added. A tailings pond is used to store the spent solution so that suspended solids settle, and the effluents can be contained without affecting the surrounding environment.

Solution Purification and Concentration

Leach solutions generally contain a large amount of impurities such as molybdenum, vanadium, selenium, iron, and copper. The concentration of pregnant solutions may range between 0.6 and 2.0 g/l U_3O_8. Solvent extraction techniques and, in some cases, ion exchange are used to purify and concentrate the U_3O_8 in the solution. The final solution, after solvent extraction, can be made to be 30 to 50 g/l. Ion exchange of clarified leach solution usually results in an elluant containing 10 to 12 g/l U_3O_8 [Reed, et al. 1979].

By using an ion exchange unit first and then a solvent extraction process, the amount of solvent required to recover the uranium may be greatly reduced. This modification has an economic advantage, by reducing solvent inventory, but may also lessen the pollution effects due to solvent entrainment and degradation.

Recommended Areas of Process Modification

After evaluating the process technology, the following areas are recommended for further research and development:

- Evaluate new leaching solutions that may offer improvement in leaching selectivity and/or efficiency.

- Evaluate solvents for improved solvent extraction of uranium salts with minimal solvent degradation.

- Divert recycle streams into additional process circuits to remove and recover trace metals.

REFERENCES

Barbour, A.K. "The Environmental Pressures on Nonferrous Metals Production", Metals and Materials, Vol. No. 1980, pp. 42-46.

Biswas, A.K. and Davenport, W.G. Extractive Metallurgy of Copper, 2nd edition, Pergamon Press 1980.

"Cyprus Bagdad: ´State of the Art´ Copper-Molybdenum Production", Engineering and Mining Journal, Vol. 181, No. 8, 1980, pp. 56-63.

Flett, D.S. "Solvent Extraction in Copper Hydrometallurgy" Trans. Instn. Min. Metall., Section C, Vol. 83, March, 1974, pp. 30-3.

Gill, C.B. Nonferrous Extractive Metallurgy, John Wiley & Sons, Inc., 1980.

Reed, A.K., Meeks, H.C., Pomeroy, S.E., and Hale, V.Q. Assessment of Environmental Aspects of Uranium Mining and Milling, EPA-600/7-76/036, Dec. 1979.

Schwitzebel, K., et al. Trace Elements Study at a Primary Copper Smelter, Volume I., EPA/600/2-78/065a, March, 1978.

U.S. EPA. Development Document for Interim Final Effluent Limitations Guidelines and Proposed New Source Performance Standards for the Primary Copper Smelting Subcategory and the Primary Copper Refining Subcategory of the Copper Segment of the Nonferrous Metals Manufacturing Point Source Category, EPA 440/1-75/032-b, February, 1975.

Wadsworth, M.E. "Review of Developments in Hydrometallurgy in 1978", Journal of Metals, Vol. 31, no. 5, 1979, pp. 12.

Wadsworth, M.E. "Review of Developments in Hydrometallurgy in 1979", Journal of Metals, Vol. 32, no. 4, 1980, pp. 27-31.

SUPPLEMENTAL REFERENCES

Agers, D.W. and DeMent, E.R. "The Evaluation of New LIX Reagents for the Extraction of Copper and Suggestions for the Design of Commercial Mixer-Settler Plants." Ingenieursblad, 41, 1972, pp. 433-4.

Anderson, S.O.S. and Reinhardt, H. "MAR-Hydrometallurgical Recovery Processes.", Paper given for the ISEC 77, Sept. 1977, pp. 798-804 of the CIM Special Vol. 21, 1979.

Ansden, M., Swietin, R., and Treilhard, D. "Selection and Design of Texas Gulf Canadian Copper Smelters and Refinery", Journal of Metals, Vol. 30, No. 7, 1978, pp. 16-26.

"Arizona's Copper Producers Rally to Fight the E.P.A. Smelters Regulations", Engineering and Mining Journal, Vol. 179, No. 4, 1978, pg. 33.

Ashbrook, A.W., Itykovich, I.J., and Sowa, W. "Losses of Organic Compounds in Solvent Extraction Processes" Paper given for the ISEC 77, Sept. 1979, pp. 781-790 of CIM Special Vol. 21, 1979.

Barner, H.E., Hulbred, G.L., and Kilumpar, I.V. "Sensitivity of LIX Plant Costs to Variations of Process Parameters", Paper given for the ISEC 77, Sept. 1977, pp. 552-560 of CIM Special Vol. 21, 1979.

Barthel, G. "Solvent Extraction Recovery of Copper from Mine and Smelter Waters", Journal of Metals, Vol. 30, No. 7, 1978, pp. 7-12.

Clarksville Zinc Plant First in the U.S. for 37 Years", Mining Magazine, Vol. 140, No. 12, 1979, pg. 551.

Cogut, B. "Morenci Clear-Air Prediction System Has Unique Potential for Controlling Air Quality", Engineering and Mining Journal, Vol. 179, No. 4, 1978, pp. 65-70.

Collins, G., Cooper, J.H., Brandy, M.R. "Designing Solvent Extraction Plants to Cut the Risk of Fires", Engineering and Mining Journal, Vol. 179, No. 12, 1978, pp. 58-62.

Costle, G.M. "Environmental Regulation of the Metals Industry", Journal of Metals, Vol. 30, No. 1, 1978, pp. 30-32.

Davidson, D.H. "In-Situ Leaching of Nonferrous Metals" Mining Congress Journal, Vol. 66, No. 7, 1980, pp. 52-57.

Finney, S.A. "A Review of Progress in the Application of Solvent Extraction for the Recovery of Uranium from Ores Treated by the South African Gold Mining Ind." Paper given to the ISEC 77, Sept. 77, pp. 567-576 of the Canadian Institute of Mining and Metallurgy, Special Volume 21, 1979.

Henrie, F. "Gearing up to Control Trace Elements During Mineral Processing", Mining Congress Journal, Vol. 66, No. 4, 1980, pp. 18-20.

Huff, R.V., Davidson, D.A., Boughman, D., and Axen, S. "Technology for In-Situ Uranium Leaching", Mining Engineering, Vol. 32, No. 2, 1980, pp. 163-164.

"Hydrometallurgy Review, 1978", Mining Engineering, Vol. 31, No. 5, 1979, pp. 527-529.

"Hydrometallurgy Review, 1979" Mining Engineering, Vol. 32, No. 5, 1980, pp. 540-542.

Isenberg, E. "Alternatives for Hazardous Waste Management in the Metals Smelting and Refining Industries, EPA/530/SW-153c.

"Jersey Mines Zinc: Plant Design and Startup", Engineering and Mining Journal, Vol. 181, No. 7, 1980, pp. 65-88.

Jones, H.R. Pollution Control in the Nonferrous Metals Industry, 1972, Noyes Data Corp., 1972.

Kordosky, G.A., MacKay, K.D., and Virnig, M.J. "A New Generation of Copper Extractant", Paper presented for Metallurgy Society of AIME, 1976.

MacDonald, B.J. and Weiss, M. "Impact of Environmental Control Expenditures of U.S. Cu, Pb, and Zn Mining and Smelting", Journal of Metals, Vol. 30, No. 1, 1978, pp 24-29.

Mackay, D., and Medir, M. "The Applicability of Solvent Extraction to Wastewater Treatment", Paper given to the ISEC 77, Sept. 1977, pp. 791-797 of the CIM Special Vol. 21, 1979.

Mackiw, V.W. "Current Trends in Chemical Metallurgy", The Canadian Journal of Chemical Engineering, Vol. 46, No. 2, 1968, pp. 3-15.

McGarr, H.J. "Solvent Extraction Stars in Making Ultrapure Copper", Chemical Engineering, Vol. 77, No. 17, Aug. 10, 1970, pp. 82-84.

Mizrahi, J., Barnes, E., and Meyer, D. "The Development of Efficient Industrial Mixer-Settlers", Paper Presented to the International Solvent Extraction Conference 1974 (ISEC 74), Sept. 1974.

Renzoni, L.S. "Extractive Metallurgy at International Nickel -- A Half Century of Progress", The Canadian Journal of Chemical Engineering, Vo. 47, No. 2, 1969, pp. 3-11.

Schultz, D.A. "Pollution Control and Energy Consumption at U.S. Copper Smelters", Journal of Metals, Vol. 30, No. 1, 1978, pp. 14-20.

Stelling, III, J.H.E. "Source Category Survey: Uranium Refining Industry", EPA-450/3-80/010, May 80.

U.S. EPA. Solution Mining of Uranium: Administrator's Guide, PB-301 173, May 79.

U.S. EPA. Heavy Metal Pollution From Spillage at Ore Smelters and Mills, EPA/600/2-77/171, Aug. 1971.

U.S. EPA. <u>Economic</u> <u>Impact</u> <u>of</u> <u>Environmental</u> <u>Regulation</u> <u>on</u> <u>the</u> <u>U.S.</u> <u>Copper</u> <u>Industry</u>, EPA/230/3-78/002, Jan. 1978.

Virnig, M.J. "Synthetic Structure, and Hydrometallurgical Properties of LIX 34", Paper given for the International Solvent Extraction Conference in 1977 (ISEC 77), Sept. 1977, pp. 535-541.

Weisenberg, I. "Design and Operating Parameters for Emission Control Studies; Kennecott, Hurley", EPA/600/2-76/036d, Feb. 1976.

Weisenberg, I. "Design and Operating Parameters for Emission Control Studies; Phelps Dodge, Morenci", EPA/600/2-76/036g, Feb. 1976.

CHAPTER 2

ELECTROPLATING

ELECTROPLATING OF COMMON METALS

Electroplating involves a series of process steps that include the preparation of the part in addition to the plating operation. Figure 2.1 is an example flow chart, of chromium plating of zinc die castings. The sequence and/or the process steps may vary from plant to plant because of the many variables involved with electroplating. Table 2.1 is a list of pollutants that typically exist in electroplating waste streams and their respective concentration ranges.

There are numerous methods of treatment for dissolved metals in effluent streams. They are [U.S. EPA, 1974]:

- Ion exchange
- Reverse osmosis
- Electrodialysis
- Chemical precipitation
- Ion flotation
- Carbon absorbtion
- Liquid-Liquid extraction

Each method has advantages and disadvantages that must be considered with respect to the specific electroplating industry, and precipitation currently has the widest application in treatment of electroplating wastewaters. Since most of these technologies are well established, similar methods can be adopted as additonal process steps to remove and recover trace metals, with the water recycled back into the process.

Recommendations for Process Modifications

The best way to reduce the amount of water needed in an electroplating process is to maintain good "housekeeping" practices to avoid spillage or

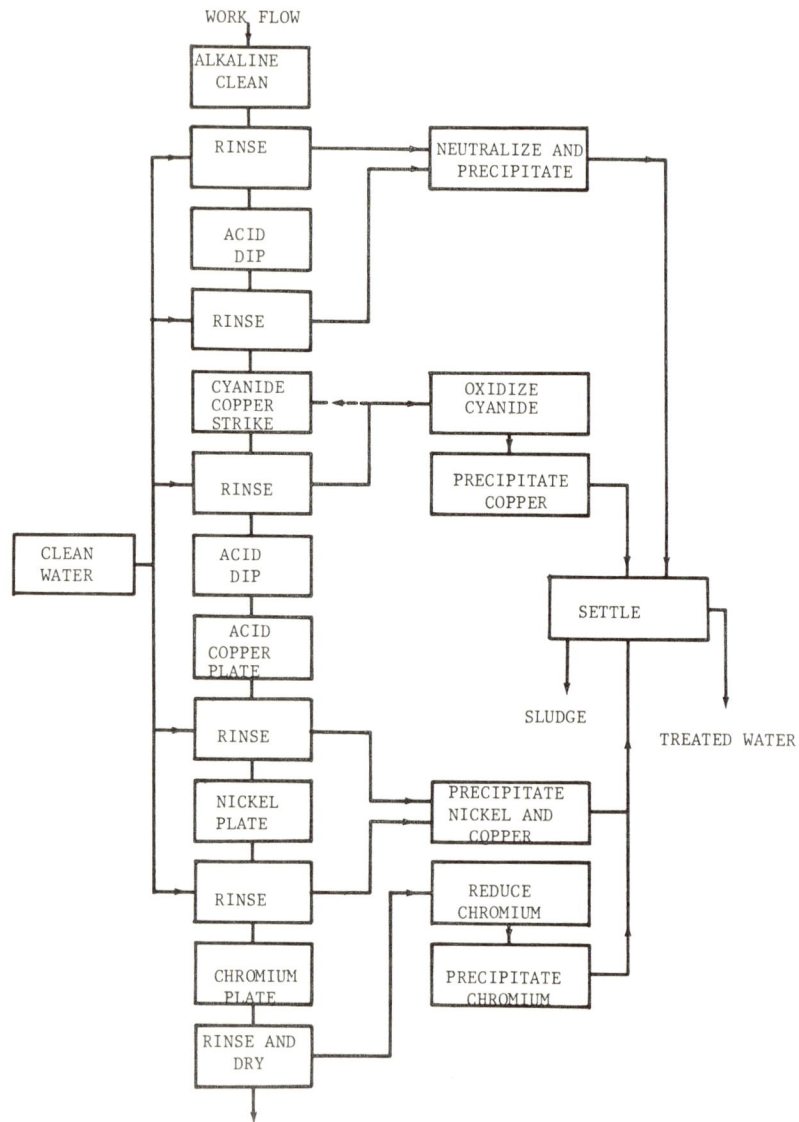

Figure 2.1. Flow Chart for Water Flow in Chromium Plating Zinc Die Castings, Decorative (U.S. EPA, April 1975).

TABLE 2.1   COMPOSITION OF RAW WASTE STREAMS FROM
            COMMON METALS PLATING [U.S. EPA, 1979]

	(mg/l)
Copper	0.032 - 272.5
Nickel	0.019 - 2954
Chromium, Total	0.888 - 525.9
Chromium Hexavalent	0.005 - 334.5
Zinc	0.112 - 252.0
Cyanide, Total	0.005 - 150.0
Cyanide, Amenable to Chlorination	0.003 - 130.0
Fluoride	0.022 - 141.7
Cadmium	0.007 - 21.60
Lead	0.663 - 25.39
Iron	0.410 - 1482
Tin	0.060 - 103.4
Phosphorus	0.020 - 144.0
Total Suspended Solids	0.100 - 9970

minimize "dragout" of treating solutions into rinse waters. Other recommendations include:

- Substitute low concentration solutions in place of high concentration baths.
- Use non-cyanide solutions in place of the cyanide treatments.
- Use counter-flow rinses.
- Add a wetting agent to rinse waters.
- Install air or ultrasonic agitation.

- Recover for reuse metals that are in effluent streams by solvent extraction.
- Recycle used rinse waters into the make-up solutions of their respective treating baths.

REFERENCES

U.S. EPA. *Development Document for Interim Final Effluent Limitations Guidelines and Proposed New Source Performance Standards for the Common and Precious Metals Segment of the Electroplating Point Source Category*, EPA/440/1-75/040, April 1975.

U.S. EPA. *Development Document for Existing Source Pretreatment Standards for the Electroplating Point Source Category*, EPA/440/1-79/003, Aug. 1979.

SUPPLEMENTAL REFERENCES

Belmont, T.V. and Cunniff, J.G. "Plating Waste Treatment", *Industrial Wastes*, Vol. 26, No. 6, 1980, pp. 14 & 27.

Elicker, L.N. *Evaporative Recovery of Chromium Plating Rinse Water*, EPA/600/2-78-127, June 1975.

Hallowell, J.B. *Assessment of Industrial Hazardous Waste Practices, Electroplating and Metal Finishing, Job Shops*, EPA/530/SW-136C, Sept. 1976.

Landrigan, R.B. and Hollowell, J.B. *Removal of Chromium from Plating Rinse Water Using Activated Carbon*, EPA/670/2-75/055, June 1975.

McDonald, C.W. *Removal of Toxic Metals from Finishing Wastewater by Solvent Extraction*. EPA/600/2-78/011, February, 1978.

Petersen, R.J. and Cobian, K.E. *New Membranes for Treating Metal Finishing Effluents by Reverse Osmosis*, EPA/600/2-76/197.

Poon, C. "Electrochemical Treatment of Plating Rinse Water", *Effluent and Water Treatment Journal*, July 1979, pp. 351-355.

Rozielle, L.T., Kopp, Jr., C.V., and Gobian, K.E. *New Membranes for Reverse Osmosis Treatment of Metal Finishing Effluents*, EPA/660/2-73/033.

U.S. EPA. *Control of Volatile Organic Emissions from Solvent Metal Cleaning.* EPA/450/2-77/022.

U.S. EPA. *Copper, Nickel, Chromium and Zinc Segment of Electroplating Development Document for Effluent Limitations Guidelines and New Source Performance Standards*, EPA/440/1-74/--3a.

U.S. EPA. *Innovative Rinse-and-Recovery System for Metal Finishing Process*, EPA/600/2-77-099.

U.S. EPA. *Second Conference on Advanced Pollution Control for Metal Finishing Industry*, EPA/8-79/014.

U.S. EPA. *Source Assessment: Solvent Evaporation-Degreasing Operations*, EPA/600/2-79/019f.

U.S. EPA. *Waste Water Treatment and Reuse in a Metal Finishing Job Shop*, EPA/670/2-74/042.

Yost, K.J. and Scarfi, A. "Factors Affecting Zinc Solubility in Electroplating Waste", *Journal of the Water Pollution Control Federation*, Vol. 51, No. 7, 1979, pp. 1878-1887.

CHAPTER 3

COAL CONVERSION PROCESSES

INTRODUCTION

The unpredictability of the international energy market and the very danger of global oil shortages in the next few decades, have necessitated a rapid expansion in the domestic energy base in the U.S. Consequently, the commercial production of synthetic fuels from the abundant reserves of coal is a major objective of the nation's energy research and development programs. Coal liquefaction and coal gasification technologies have received renewed interest in this regard. Economic viability and environmental impact will be the limiting factors in the commercialization of such processes.

LIQUEFACTION

All coal liquefaction processes produce liquids form coal, by yielding a material having a higher hydrogen content than coal. Fuel oils contain about 9 percent hydrogen and gasoline 14 percent as compared to roughly 5 percent hydrogen in raw coal. Currently, some twenty-odd liquefaction processes are in various stages of development by industry and federal agencies. Coal liquefaction technologies can be categorized under hydrogenation, pyrolysis and hydrocarbonization, and catlytic synthesis. Of these, hydrogenation is the most advanced [Koradek and Patel, 1978].

In this chapter, the environmental problems associated with the following four coal liquefaction technologies will be discussed:

1. Solvent refined coal, (SRC)
2. Synthoil
3. H-Coal
4. Exxon donor solvent

Solvent Refined Coal Liquefaction

A fully integrated SRC liquefaction system flow scheme [Shield, et al. 1979] is shown in Figure 3.1. Raw coal from the coal storage facilities is sent to the coal pretreatment operation where it is sized, dried, and mixed with reactor product slurry recycled from the gas separation processes. The resulting feed slurry is combined with recycled hydrogen from the hydrogen/hydrocarbon recovery process and makeup hydrogen. This hydrogen-rich slurry is pumped through a preheater to the liquefaction reactor or dissolver. Exothermic hydrogenation reactions initiated in the preheater continue in the dissolver, which typically operates between $435^{\circ}C - 470^{\circ}C$. The reactor product slurry is sent to the gas separation processes where the gaseous products are removed. Auxiliary processes separate these gases into components including recycled hydrogen, SNG, LNG, and sulfur salts. The sulfur species are further converted to by-product elemental sulfur. Part of the separated slurry from gas separation is recycled to the coal pretreatment operation. The remainder of the slurry is sent to the fractionator. The fractionator generates three streams; a light distillate which is hydrotreated to form naphtha and fuel oil products, liquid SRC, the primary product, and a bottom stream which is sent to solids/liquids separation processes. The vacuum distillation unit in the solids/liquids separation recovers additional SRC liquid from the fractionator bottoms, yielding a residue of high mineral matter content. Part of this residue is gasified to produce makeup hydrogen.

Particulate Emissions

Fugitive emissions (i.e., particulates) will be emitted from the following sources: coal storage piles, coal reclaiming and crushing, coal receiving, dryer stack gas, and ash from stream generation. Trace elements of fugitive emissions from coal preparation represent a potential health hazard [Hopkins, et al. 1978]. Trace element concentrations in particulates escaping from treated stack gas are enriched in zinc, copper, zircronium, molybdenum and selenium. The trace element concentration of coal dust is expected to be similar to that of the parent coal, although certain trace elements may concentrate in the smaller size range.

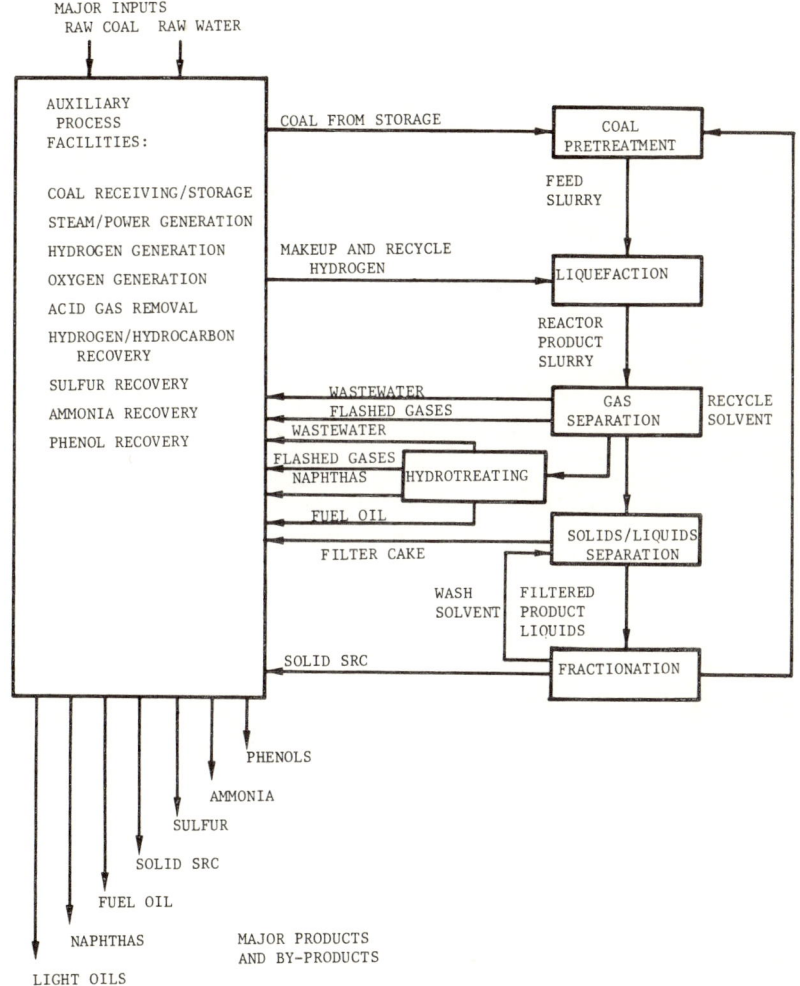

Figure 3.1. Flow Sheet for An Integrated Liquefaction Process

Dust typically 1 to 100 μ in size) generated from coal receiving, storage, reclaiming and crushing is estimated to be about 24 tons/day, for a 20,000 ton/day SRC II plant [Rogoshewski, et al. 1978]. Table 3.1 lists an analysis of some trace elements in dust from coal preparation after treatment with a wet scrubber [Hopkins, et al. 1978].

Suggested control alternatives for coal dust include polymer spraying, enclosed storage, and water spraying. Of these, enclosed storage involves $6-8 million capital investment for a 10,000 ton coal pile [Rogoshewski, et al. 1978]. Cyclone and baghouse filters appear to be the most effective methods for controlling coal dust. Due to the small particle range, magnetic and electrostatic filters are needed.

Table 3.1 Approximate Trace Element Analysis of Coal Pretreatment Dust after Wet Scrubbing

Element	g/day
Aluminum	8,910
Arsenic	3.9
Chromium	13.0
Nickel	15.0

Coal Pile Drainage

Coal pile runoff resulting from rainfall, etc., may contain oxidation products of metallic sulfides. This runoff may also be acidic, with relatively high concentrations of suspended and dissolved solids [Rogoshewski, et al. 1978]. The elements in this runoff include calcium, iron, aluminum and manganese. Wastewater generated in coal preparation is routed to a thickener where particulates are removed as underflow. This underflow contains about 35 percent suspended solids; these solids being expected to contain the same elements as those of the original coal. Thickener underflow is mixed with coal pile runoff and sent to a tailings pond.

Gasifier Slag and Fly Ash

One of the large volume solids to be disposed of in a SRC plant is the gasifier slag and quenched fly ash. Current design specifications for the treatment of slag require that it be crushed, slurried with water, and de-ashed. The fly ash from the quencher is flashed down, neutralized with lime and thickened [Hopkins, et al. 1978]. It is anticipated that this waste can be disposed of in strip mines. The leachability of this waste is one aspect which needs further investigation.

Residue from Solids/Liquid Separation

Solid residue from solids/liquid fractionator bottoms cannot be completely utilized in hydrogen production, due to its high ash content ($\simeq 64$ percent). Efforts to dissolve the solid in dilute acid did not produce any leacheate. Table 3.2 gives the composition of some of the elements in this mineral residue [Hopkins, et al. 1978], which totals about 3,700 MG/day for a 10,000 ton/day SRC-II liquefaction plant.

Table 3.2 Estimated Trace Elements Composition of the SRC Liquefaction Residue

Element	Concentration ppm
Arsenic	24.9
Barium	579
Calcium	33,323
Iron	116,760
Luterium	2,050
Nickel	126
Sodium	1,155
Zinc	1,938

The extraction and recovery of metals from this residue is another area that requires further evaluation.

Spent Catalyst

The catalyst used in the shift reaction in hydrogen generation has been estimated at about 135 m^3 [Hopkins, et al. 1978]. Trace elements, sulfur compounds and heavy hydrocarbons may be adsorbed on the catalyst. The estimated lifetime of a cobalt molybdate catalyst is three years. Regeneration of spent catalyst may not be possible, due to the presence of trace metals and sulfur as well as possible sintering of the catalyst. It is suggested that valuable metals be extracted from the spent catalyst at an off-site facility. Similar facilities may be needed for spent sulfur guard and methanation catalysts.

Tail Gas From Acid Gas Removal

Acid gas from the gas purification module and gas from hydrogen production are routed to a Stretford unit, where H_2S [Hopkins, et al. 1978] is recovered as elemental sulfur. The Stretford process has an efficiency greater than 99.5 percent in sulfur recovery and can reduce H_2S concentrations to less than 10ppm. However, tail gas treatment may be needed before the spent gas can be vented. The Stretford solution consists of sodium metavanadate, sodium anthraquinone disulfonate (ADA), sodium carbonate and sodium bicarbonate, in water. The Stretford solution purge stream has a total salt concentration of 10-25 percent. Using a high temperature hydrolysis technique, vanadium (as solid) and sodium carbonate, sulfate and sulfite can be recovered as solids. HCN is completely converted to CO_2, H_2O and N_2 [Rogoshewski, et al. 1978]. The Stretford tail gas contains about 42.7 percent CO_2 and 5,500 ppm hydrocarbons (as C_2H_6). Direct flame incineration and carbon adsorption with incineration (AdSox) is recommended as tail gas treatment for hydrocarbons. Further scrubbing might be needed to remove CO_2 in the tail gas. It is suggested that the Stretford solution be modified to reduce CO_2 emissions in the tail gas. This is one area which necessitates further research on adsorption phenomena of SO_x and CO_2 in various solvents.

NO_x Removal

The reported sources of NO_x emissions within the SRC plant include [Hopkins, et al. 1978]:

    Steam generation   8.4 Mg/day
    Stretford effluent gas   0.005 Mg/day

The reduction in NO_x emissions during steam generation from reducing air flow rates may not be as effective or damage-proof as other modification techniques. By supplying substoichiometric quantities of primary air to burners, a 40 to 50 percent reduction in NO_x emissions has been observed. Another useful method of reducing NO_x is achieved by recirculation of coal flue gas, which lowers peak flame temperature [Rogoshewski, et al. 1978].

Table 3.3 is a summary of some of the pollution problems associated with SRC liquefaction and suggested control alternatives.

Recommended Areas for Pollution Control Research

After evaluating the process technology, the following areas are recommended for further research and development:

- The applicability of electrostatic and magnetic filters to control emissions of coal dust particles in the submicron range.

- Leachability of gasifier slag and fly ash to determine treatment needs and/or disposal limitations.

- Extraction of possible toxic and/or valuable metals from solids/liquids separation residue.

- Extraction of valuable metals (Ni, Co, Mo, etc.) from spent shift and hydrogen generation catalysts.

- Studies on the absorption of SO_x and CO_2 in Stretford process leading to process modification to reduce CO_2 emissions in tail gas.

Table 3-3 Pollutants and Control Options in SRC Liquefaction Process

Process	Pollutants	Sources in Process	Nature of Pollutants	Pollutant Control Strategy
Coal Pretreatment	Coal Dust	Storage, handling Sizing	Similar Original Coal	Cyclone, wet scrubbing, polymer Coating, Electrostatic and Magnetic Filters
Coal Pretreatment	Thickener Underflow	Coal Pile Run Off	Minerals and Suspended Solids	Extraction and Disposal of Tailings.
Gasifier	Slag, Fly Ash	--	Inorganics, Metals and Organics	Leaching of Metals and Strip Mining.
Solids/ Liquids Separation	Mineral Residue	Fractionator Bottoms	Inorganics, Metals and Organics	Extraction of (Toxic) Metals
Hydrogen Generation	Spent Catalyst	Reactor	--	Extraction of Co, Mo, Ni
Acid gas Removal	Tail gas	Stretford Process	Hydrocarbons and CO_2 ($\approx 43\%$)	Direct Flame Incineration, Modification of Stretford process to remove CO_2
Steam Generation	NO_x Emissions	--	NO_x gases	Substoichiometric supply of air, Recirculation of Flue Gas

- Modification of steam generation operation to reduce NO_x emissions.

GASIFICATION

In this section, the pollution problems and control alternatives for the following four High-Btu coal gasification technologies will be discussed.

1. Lurgi
2. Hy-Gas
3. Bi-Gas
4. Koppers-Totzek

Lurgi Coal Gasification

The conversion of coal to SNG involves the reaction of coal with steam and oxygen in a gasifier, with subsequent gas processing to (a) adjust the H:CO ratio by water-gas shift reaction, (b) remove acidic components, and (c) catalytic methanation. The corresponding chemical reactions are:

$$6(C+H) + \frac{5}{2}O_2 + 3H_2O \rightarrow 4H_2 + 2CO + 3CO_2 + CH_4$$
Lurgi gasification (3.1)

$$CO + H_2O \rightarrow CO_2 + H_2$$ Water-gas shift (3.2)

$$3H_2 + CO \rightarrow CH_4 + H_2O$$ Methanation (3.3)

The four major processes in the Lurgi gasification are discussed below. A flow sheet of the process is presented in Figure 3.2.

Coal Preparation

Coal pretreatment in the Lurgi process generally consists of only crushing and screening the coal, to produce 3-35 mm particles. This larger size range, compared to certain other gasification processes, decreases fugitive emissions.

Coal Gasification

The coal gasifier is operated at a pressure of about 25 to 35 atm, receives coal through a feed hopper at the top, and discharges ash through the

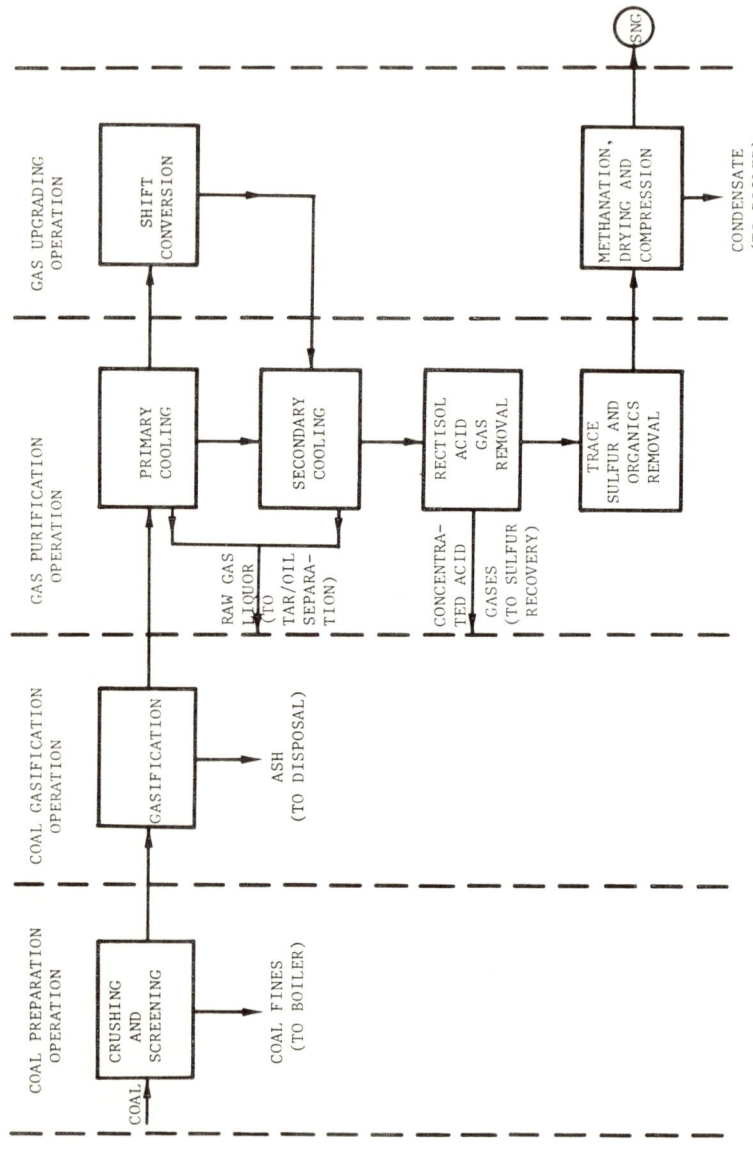

Figure 3.2. Lurgi SNG Process.

bottom. Oxygen and steam enter at the bottom of the gasifier and the product gas exits near the top. On a dry basis, the product gas contains about 4 percent H_2, 3 percent CO_2, 18 percent CO and 10 percent CH_4, as well as higher molecular weight hydrocarbons, reduced sulfur and nitrogen compounds.

Gas Purification

Gas purification consists of the removal of condensables by cooling, Rectisol treatment for the removal of bulk CO_2 and reduced sulfur compounds, and removal of trace sulfur using "methanation guards." The condensates produced are sent to tar oil separation units, and by-product recovery. The Rectisol process uses cold methanol to absorb acid gases under pressure. The used solvent is regenerated by stepwise depressurization and heating. Methanation guards are beds of solid adsorbent (e.g. ZnO). The exhausted beds are usually discarded rather than regenerated.

Gas Upgrading

Cobalt molybdate is used as a shift catalyst to obtain a 3:1 H_2:CO ratio. Nickel based materials are usually used as methanation catalysts. Both the spent shift and methanation catalysts are solid wastes requiring treatment for metal recovery and/or disposal.

Gasifier and Boiler Ash

Wet ash from the gasifier and boiler ash quench systems is the largest volume solid waste stream in an SNG plant [Ghassemi, et al. 1979]. A 250 x 10^6 Scf/day Lurgi SNG plant using a coal containing 15 percent ash is expected to generate about 5400 ton/day wet ash. This ash contains leachable inorganic and organic materials. These constituents can be possible sources of groundwater contamination. An alternative disposal scheme would involve stabilization to convert the wastes into a chemical form that is more resistant to leaching in the ultimate disposal site. Table 3.4 presents typical leaching study results for wastes stabilized by the Chem-Fix process [Conner, 1974].

Table 3.4 Laboratory Leaching Results of Chem-Fixed Refinery Wastes

Element	Concentration in Raw Sludge, ppm	Concentration in approx. 200 ml Leachate water after Chem-Fix, ppm
Chromium	43.5	< 0.1
Iron	1310	< 0.1
Zinc	88.0	< 0.1
Nickel	8.9	< 0.1

The leachability of the ash produced in Lurgi SNG is an area which requires further investigation.

Spent Guard Catalyst

Zinc oxide beds are used for removing sulfur after the gas purification step has removed all but a few parts per million. Water vapor content is critical, because liquid water can completely degrade the ZnO bed [Ghassemi, et al. 1979]. The sulfur loading capacity of the ZnO catalyst increases from $\simeq 5$ wt% at $273°K$ to $\simeq 20$ wt% at $673°K$. However, the maximum recommeded loading is only 3 wt% when the desired exit gas specification is 0.02 ppm H_2S [Drano Corp., 1978]. Spent methanation guard material will consist primarily of zinc sulfide and unreacted zinc oxide. Operating data on the quantity and composition of spent methanation guard catalyst are needed to evaluate the disposal and/or reclamation of the spent catalyst [Ghassemi, et al. 1979].

Spent Shift/Methanation Catalysts

Catalysts used for shift and methanation require periodic replacement; the spent catalysts consistute solid wastes. Gross composition of a spent catalyst is not expected to be dramatically different from that of a fresh catalyst, although accumulation of carbon, sulfur, and metallic elements is to be expected. Data is needed on the characteristics of spent Lurgi SNG catalysts. Cobalt-molybdate is used as a shift catalyst. Table 3.5 presents pilot plant data on spent Harshaw Ni-0104-T-1/4 catalyst [Leppin, 1977].

Table 3.5 Spent Harshaw Nickel Catalysts Analysis

	Typical Fresh Catalyst	Bottom of First Stage Methanator	Top of Second Stage Methanator
Sulfur, %wt.	0.15	3.7	0.16
Carbon, %wt.	3.4	3.4	4.5
Nickel, %wt.	60.0	52.0	61.0

Spent nickel methanation catalyst is as active as zinc for trace sulfur removal, and can be used as a sulfur guard catalyst [Ghassemi, et al. 1978]. Sequential interchange of the first and second stage methanators is also recommended, as it is apparent from Table 3.5 that the catalyst at the top of the second stage methanator is almost fresh. In the end, however, spent methanation catalyst constitutes a solid waste. The large weight percent of nickel on the catalyst would be an economic incentive for reclaiming that metal from the spent catalyst. Data on the properties of spent nickel methanation catalysts and the applicability of metal extraction for reclamation are two areas that deserve further investigation. Further, the spent catalysts, although of small quantity, are of special concern due to their content of potentially toxic metals (Ni, Co, Mo), coal-derived organics (metal carbonyls for example), and trace elements. The fixed-bed methanation/shift reactor system is reported to have the advantages of (1) operation at conditions removed from carbon formation, thereby increasing catalyst life, (2) reduction in catalyst sensitivity to sulfur, and (3) need for CO_2 removal from a reduced volume of gas. However, this has not yet been commercially demonstrated [Gassemi, et al. 1978].

Ash Quench Slurry

An ash quench slurry results when process waters are used to cool and transport gasifier ash to a settling unit or disposal site. No operating data are available for ash quench slurry characteristics [Ghassemi, et al. 1979]. Laboratory data indicate that this slurry can contain up to 20 g/L of dissolved solids, with the dominant ions being sodium, potassium, calcium, and sulfate [Griffin, et al. 1977]. An increase in the solubility of iron, manganese, cadmium, and aluminum is also expected with decreasing pH. In addition, the concentration of certain hazardous elements (e.g. arsenic, chromium and copper) can reach levels which warrant control of the ash slurry discharge. Characterization of ash slurry water is necessary to evaluate pollution control alternatives such as coagulation and flocculation.

Phenol Recovery

Gas liquor from tar-oil recovery contains a high concentration of phenols, up to 4200 mg/L [Ghassemi, et al. 1979). The Phenosolvan process, using butyl acetate as solvent, is used to extract these phenols. Most of the available data on the performance of this solvent extraction process are for one unit in Salsbury, South Africa. Solvents suggested for the Phenosolvan process include butyl acetate, isopropyl ether and light aromatic oil [Wurm, 1969; Earhart, et al. 1977]. For butyl acetate, the following distribution coefficients for various phenolic compounds have been reported [Earhart, et al. 1977].

Table 3.6 Distribution Coefficients for Various Phenols in Butyl Acetate at 300°K

Compound	K*
Phenol	65
3,5 Xylenol	540
Pyrocatechol	13
Resorcinol	10

$K_D = \dfrac{\text{wt. fraction in solvent phase}}{\text{wt. fraction in a aq. phase}}$, measured at high dilution

The solvents can remove only limited amounts of non-phenolic organics. A better characterization of the available solvents for phenol extraction, and their removal efficiencies for various phenols, is needed. Further development of solvent extraction systems to remove non-phenolic organics is also suggested.

Removal of Sulfur Compounds from the Rectisol Process

In terms of total volume and content of H S and other reduced sulfur compounds, concentrated acid gas from the Rectisol process is the most important gaseous waste stream in a Lurgi SNG facility. Some of the control options for the concentrated acid gas stream are shown in Table 3.7 [Ghassemi, et al. 1979].

Table 3.7 Control Options for the Concentrated Acid Gas Stream

Control Options	Comments
1. Claus Plant Sulfur Recovery	High Concentration of H_2S in tail gas; feed gas H_2S enrichment and Hydrocarbon removal are needed
2. Claus Plant Sulfur Recovery and tail gas treatment	Not highly effective at high levels of CO_2 in feed, applicable only for streams containing 5-15 percent H_2S
3. Claus Plant Sulfur recovery and SO_2 control/recovery	Reasonable option when feed gases contain more than 5-15 percent H_2S
4. Stretford Sulfur Recovery	Inapplicable to was gases containing greater than 15 percent H_2S; not economical at high CO_2 levels; discharge may contain high COS and HC levels

Table 3-8. Pollutants and Control Options in Lurgi SNG Process

Process	Pollutants	Sources in process	Nature of Pollutant	Pollutant control Strategy
Coal Gasification	ash	--	leachable inorganic, organic materials	stabilization of ash to decrease or prevent leach ability
Gas Upgrading	Spent methanation guard catalyst	Methanation guard reactor	ZnO with trace organics, sulfur and metals	Reclamation by extraction
Shift/Methanation	Spent catalysts	Reactors	Co, Mo, Ni, with trace organics, sulfur	Spent methanation catalyst as methanation guard, extraction of toxic metals from spent catalyst
Coal Gasification	Ash quench slurry	Thickener overflow	Dissolved inorganics, hazardous metals (As, Ni)	Coagulation and flocculation to remove dissolved salts
Tar/oil separation	Phenols in gas liquor	--	Nonphenolic organics	Better characterization of solvents to modify/develop extraction systems
Acid gas removal	Sulfur compounds	Rectisol tail gas	SO_2, H_2S, CO_2	Claus sulfur recovery and SO_2 tail gas treatment, Stretford sulfur recovery

COAL CONVERSION   47

The determination of the best option for the management of a specific sulfur-bearing gas stream must be made on a case-by-case basis, due to restrictions on feed compositions and economics for different processes. Some of the above options have not appeared in commercial design [Ghassemi, et al. 1979] due to the lack of engineering data. It appears that research and development to provide operational data for various sulfur recovery, SO_2 and tail gas treatment processes is needed.

Table 3.8 summarizes some of the pollution problems associated with the Lurgi SNG process, and suggested control alternatives.

Recommended Areas for Pollution Control Research

After evaluating the process technology, the following areas are recommended for further research and development:

- Stabilization of gasifier and boiler ash to decrease or prevent leachability.

- Operating data and composition on spent methanation guard, shift and methanation catalyst to allow evaluation of possible extraction and recovery of valuable and toxic metals.

- Studies on the use of spent methanation catalyst as sulfur guard.

- Development of solvent extraction systems to remove non-phenolic organics, and determination of distribution coefficients for existing solvents.

- Engineering data on various sulfur recovery, SO_2 and tail gas pretreatment processes.

REFERENCES

Connor, J.R. *Disposl of Liquid Wastes by Chemical Fixation*, Waste Age, Sept. 1974, pp. 26-45.

Drano Corp. *Handbook of Gasifiers and Gas Treatment Systems*, ERDA No. FE-1772-11, Pittsburgh, PA, 1976.

Earhart, J.P., et al. Recovery of Organic Pollutants Via Solvent Extraction, Chem. Eng. Progr., May 1977, p. 67.

Ghassemi, M., et al. Environmental Assessment Data Based for High-Btu Gasification Technology: Volume II. Appendices A, B, and C, EPA-600/7/78-186b, Prepared for U.S. EPA by TRW Environmental Eng. Div., Redondo, Beach, CA, Sept. 1978.

Ghassemi, M., et al. Environmental Assessment Report: Lurgi Coal Gasification Systems for SNG, EPA-600/7-79-120, Prepared for U.S. EPA by TRW Environmental Eng. Div., Redondo Beach, CA, May 1979.

Griffin, R.A., et al. Solubility and Toxicity of Potential Pollutants in Solid Coal Waste, Presented at EPA Symposium on the Environmental Aspects of Fuel Conversion Technology, Sept. 1977.

Hopkins, H.T., et al. SRC-Sit-Specific Pollutant Evaluation: Vol. I, Discussion, EPA-600/7-78-223a, Prepared for U.S. EPA. by Hittmann Associates, Inc., Columbia, Md., Nov. 1978.

Koradek, C.S. and Patel, S.S. Environmental Assessment Data Base for Coal Liquefaction Technology: Vol. I. Systems for 14 Liquefaction Processes, EPA-600/7/78-184a, Prepared for U.S. EPA. by Hittmann Associates, Inc., Columbia, Md., Sept. 1978.

Leppin, D. Ninth Synthetic Pipeline Gas Symposium, Chicago, IL, Oct. 31 - Nov. 2, 1977.

Rogeshewski, P.J., et al. Standards of Practice Manual for the Solvent Refined Coal Processes, EPA-600/7-78-091, Prepared for U.S. EPA. by Hittmann Associates, Inc., Columbia, Md., June, 1978.

Shields, K.J., et al. Environmental Assessment Report: Solvent Refined Coal (SRC) Systems, EPA-600/7-79-146, Prepared for U.S. EPA. by Hittmann Associates, Inc., Colubmia, Md., June 1979.

Wurm, H.J. *Treatment of Phenolic Wastes*, Eng. Bull. Purdue Univ., Eng. Ext. Serv., 132 (II), 1054-73, 1969.

SUPPLEMENTAL REFERENCES

U.S. EPA. *Sasol: South Africa's Oil to Coal Story--Background for Environmental Assessment*, EPA-600/8-80-002, Jan. 1980.

U.S. EPA. *Air Emissions from Combustion of Solvent Refined Coal*, EPA-600/7/79-004, Jan. 1979.

U.S. EPA. *Coal Processing Technology: Environmental Impact of Synthetic Fuels Development*, Chem. Eng. Progr., 1975, 6.

U.S. EPA. *Control Technologies for Particulate and Tar Emissions from Coal Converters*, EPA-600/7-79-170, July 1979.

U.S. EPA. *Effects of Combustion Modifications for NO Control on Utility Boiler Efficiency and Combustion Stability*, EPA-600/2-79-190, Sept. 1977.

U.S. EPA. *Engineering Evaluation of Control Technology for the H-Coal and Exxon Donor Solvent Processes*, EPA-600/7-79-168, July 1979.

U.S. EPA. *Environmental Assessment of Coal Liquefaction: Annual Report*, EPA-600/7-78-019, Feb. 1978.

U.S. EPA. *Environmental Assessment of High-Btu Gasification: Annual Report*, EPA-600/7-78-025, Feb. 1978.

U.S. EPA. *Environmental Assessment Data Base Coal Liquefaction Technology: Vol. II. Synthoil, H-Coal, and Exxon Donor Solvent Processes*, EPA-600/7-78-184B, Sept. 1978.

U.S. EPA. *Environmental Assessment Data Base for High-Btu Gasification Technology: Volume I. Technical Discussion*, EPA-600/7-78-186A, Sept., 1978.

U.S. EPA. Evaluation of Background Data Relating to New Source Performance Standards for LURGI Gasification, EPA-600/7/77-057, June 1977.

U.S. EPA. Evaluation of Pollution Control in Fossil Fuel Conversion Processes -- Liquefaction: Section 2 - SRC Process, EPA-650/2-74-009F.

U.S. EPA. Evaluation of Pollution Control in Fossil Fuel Conversion Processes -- Gasification,
Section 1 - Koppers-Totzek Process-EPA-650/2-74-009A
Section 3 - Lurgi Process - EPA-650/2-74-009C
Section 5 - Bi gas Process - EPA-650/2-74-009G
Section 6 - Hygas Process - EPA-650/2-74-009H

U.S. EPA. Mechanism and Kinetics of the Formation of NO and other Combustion Products -- Phase II Modified Combustion, EPA-600/7/76-009B, Aug. 1976.

U.S. EPA. Pollutants from Synthetic Fuels Production: Sampling and Analysis Methods for Coal Gasification, EPA-600/7-79-201, Aug. 1979.

U.S. EPA. Sulfur and Nitrogen Balances in The Solvent Refined Coal Processes (Task I), EPA-650/2-75-011, Jan. 1975.

U.S. EPA. Sulfur Retention in Coal Ash, EPA-600/7-78/1538, Nov. 1978.

U.S. EPA. Symposium Proceedings: Environmental Aspects of Fuel Conversion Technology, II, Dec. 1975, EPA-600/2-76-194, June 1976.

U.S. EPA. Symposium Proceedings: Environmental Aspects of Fuel Conversion Technology, III, Sept. 1977, EPA-600/7/78-063, April 1978.

U.S. EPA. Symposium Proceedings: Environmental Aspects of Fuel Conversion Technology, IV, April 1979, EPA-600/7-79-217, Sept. 1979.

U.S. EPA. The Solubility of Acid Gases, in Methanol, EPA-600/7-79-097, April 1979.

CHAPTER 4

EXPLOSIVES INDUSTRY

INTRODUCTION

The processes involved in manufacturing the high explosives are discussed in this chapter, and will provide an insight in reducing the pollutants arising from these activities by means of process modification. Included are the two most produced high explosives; trinitrotoluene (TNT) and nitrocellulose, which are generated by nitration processes to yield these nitrocompounds.

The explosive industry, as a whole, can be grouped into four categories:

1. Manufacturing of explosives

2. Manufacturing of propellants

3. Formulation [load, assemble and pack (LAP)]

4. Manufacturing of initiating compounds

Since both of the compounds discussed in this chapter require large amounts of nitric acid for production, the process for producing nitric acid is also presented. Process descriptions given her do not include actual formulation (load, assemble, and pack). Figure 4.1 is an overview flow chart of the explosive industry.

NITRIC ACID PRODUCTION

Figure 4.2 demonstrates the role of nitric acid in the explosive industry. The process given here is not the only process available for nitric acid production. As is suggested later in this section, other systems for this process must be closely examined.

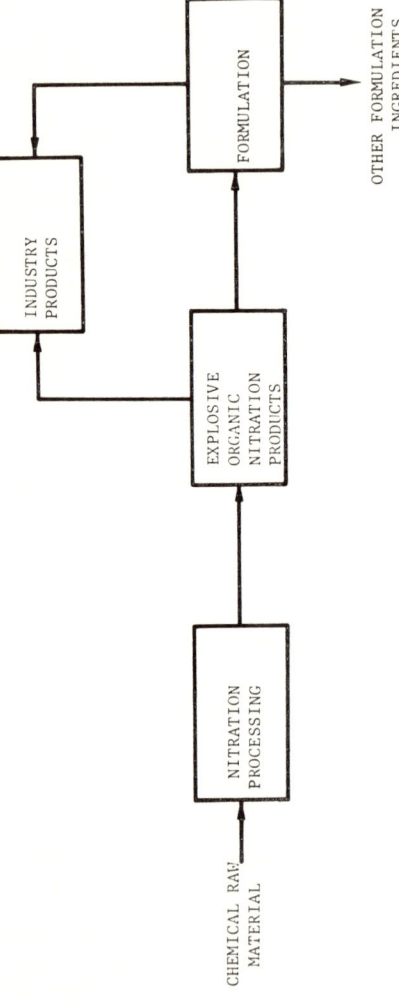

Figure 4.1. Processes in the Explosive Industry (Koradek and Patel, 1978).

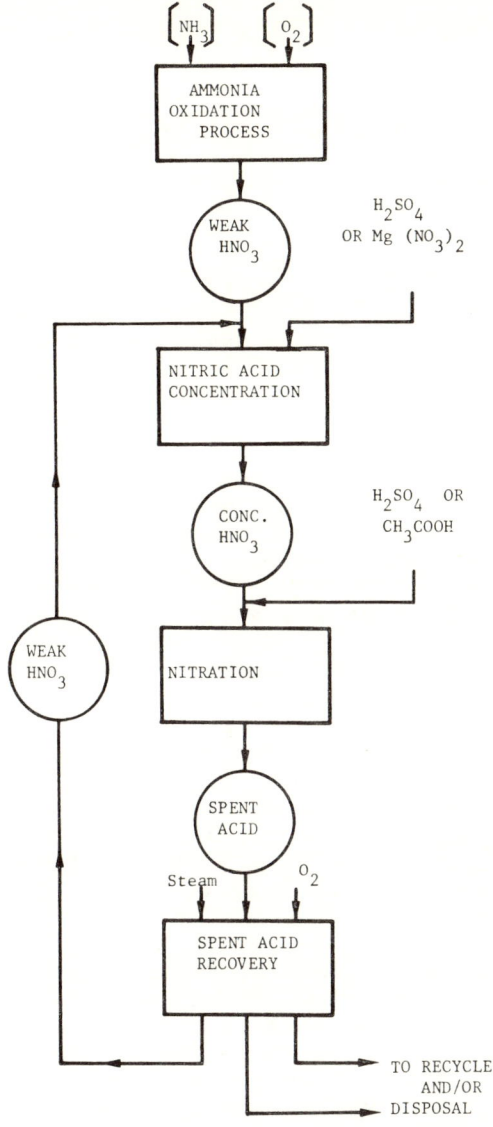

Figure 4.2. Flow Chart for Nitric Acid Production (Hudak and Parsons, 1977).

Ammonia Oxidation Process

In this process, anhydrous ammonia is vaporized and mixed with preheated air, and combusted under pressure in the presence of a catalyst, to produce nitric oxide. Nitric oxide is further oxidized by excess air to nitrogen dioxide and its dimmer (N_2O_4). Typically the catalyst is platinum-rhubidium or platinum-palladium-mercury. The bed operates at 800-919 °C and 120 psia. The equilibrium mixture of NO_2 and its dimer are adsorbed in a water cooled absorption tower to form weak (60-65 percent HNO_3) nitric acid [Hudak and Parsons, 1977; Patterson, et al. 1976].

The main source of pollution from the ammonia oxidation process (AOP) is the tail gas from the absorption tower, which produces approximately 2.5 g NO_x/kg HNO_3 for the high pressure process. The conversion is at least 95 percent of theoretical, with an ammonia consumption of 0.4 kg/kg HNO_3 [Hudak and Parsons, 1977]. In addition, the tail gas contains N_2 [Patterson, et al. 1976].

Quantitative reduction in pollutants can be achieved as a direct result of an increase in HNO_3 production. The first strategy in increasing HNO_3 output is a more efficient design for the catalytic bed. Maximum yield of nitrogen oxides can be attained by mixing the gases via a filter near the inlet before contacting the catalytic gauze, and by selection of an optimal operating temperature [Smith and Lockyer, 1979; Dorfman, et al. 1978]. On the other hand, the mechanism of the surface catalysis plays an important role in NO generation. For small concentrations of NH_3 on the catalyst surface, the conversion of NH_3 into NO is almost complete. The yield of NO can tend to decrease, since NO is decomposed to N_2 [Atroschenko, et al. 1979]. This decrease is a result of poor exchange between catalyst surface and the gas bulk, and subsequent accumulation of NO on the catalyst surface. Based on this argument, the yield of NO in the bed can be optimized by increasing the rate of mass transfer from the bulk gas to the catalyst surface. Furthermore, the degree of dissociation of NO increases linearly with time [Zhidkov, et al. 1979]. Thus, knowledge of the NO dissociation reaction and NO production rate will be needed in maximizing NO output.

It is also of value to examine the operating parameters for the absorption tower. The rate of nitric acid production can be improved by controlling the temperature of the absorber. Dorfman, et al. (1978) accomplished a 4 percent in HNO_3 output by adjusting the H_2O consumption, thus controlling the temperature.

Nitric Acid Concentration

The nitric acid concentration process is a continuous operation in which nitric acid from the AOP is mixed with concentrated H_2SO_4 and fed to the distillation tower with steam. The sulfuric acid combines with free water while (98-99 percent) HNO vapors [Hudak and Parsons, 1977] form an overhead stream. The nitric acid vapors, contaminated with small amounts of NO_x and O_2 from HNO_3 dissociation, pass to a bleacher and condensor. The HNO vapors condense as 95-99 percent HNO_3, while NO_x an O_2 pass to the absorber column for conversion to, and recovery of, additional weak nitric acid. This weak nitric acid is recycled to the dehydrating unit. The bottom product, consisting of approximately 68 percent H_2SO_4 [Hudak and Parsons, 1977], is recovered and sent to a concentration unit for reprocessing.

The principal source of emissions, 0.1 - 2.5 g NO_x/Kg HNO produced [Hudak and Parsons, 1977], is from the absorber tail gas. The NO_x content of the tail gas is affected by several variables: insufficient air supply to the absorber, high temperatures in the absorber, and internal leakage which permits gases from the AOP to enter the absorber [Hudak and Parsons, 1977].

In considering process modications which may lead to abatement of pollutants, it is advantageous to consider other alternative processes for production of nitric acid. One newly developed process involves concentrating nitric acid by surpassing the azeotrope. In this process, the gases leaving the AOP oxidation step are enriched in NO_2 and then absorbed in azeotropic acid to produce a super azeotrope mixture (80 percent HNO_3). This mixture is then easily distilled. The exiting gases from this particular process contain 300 ppm of NO_x [Marzo and Marzo, 1980]. The advantage of this process, as opposed to the conventional, oxidation absorption operation, should be considered.

Spent Acid Recovery

Spent acid from the various nitration processes flow into the top of a denitrating tower. The HNO_3 and NO_x are stripped from the spent acid by steam. The bottom product contains H_2SO_4, which is sent to a H_2SO_4 concentrator. Sulfuric acid (93 percent) from the concentrator is a by-product of most nitration processes [Patterson, et al. 1976].

Wastewaters from the spent acid recovery unit are characterized by high concentrations of suspended solids and small quantities of nitrogen salts. The solids from these processes, called nitrobodies, must be removed from the spent acid for pollution control.

Table 4.1 summarizes some of the pollution problems associated with nitric acid production and suggests control alternatives.

Recommended Areas for Pollution Control Research in Nitric Acid Production

After evaluating the process technology, the following areas are recommended for further research and development:

- Studies on kinetics and mass transfer for heterogeneous catalysis to improve the catalytic bed design, which will reduce NO dissociation and increase NH_3 conversion.

- Studies on heat and mass transfer with reaction in absorption towers to optimize HNO_3 production, by determining the optimal parameters for adsorption tower operation.

- Consideration of nitric acid production by means of a super azeotropic mixture which may be useful in modification of the present process.

TNT PRODUCTION

Trinitrotoluene (TNT) is the most extensively produced military high explosive. Production of TNT during 1969-71 exceeded 22,000 tons/month, more than any other high explosive compound [Patterson,

EXPLOSIVES 57

Table 4.1 Pollutants and Control Options in Nitric Acid Production Process

Process	Pollutants	Source in Process	Nature of Pollutants	Pollutant Control Strategy
Ammonia Oxydation Process	NO_x, N_2	Tail gas from absorption tower	Inorganic gases	Adjustment of operating temperature for maximum HNO_3 output. Mixing of gases at the inlet of catalytic bed. Improvement of mass transfer rate between the catalyst and the bulk gas.
Nitric acid concentration	NO_x	Absorption tail gas	Inorganic gases	Provide sufficient air supply. Prevention of leakage from AOP.
Spent acid recovery	Nitrogen salts nitrobodies	From nitration processes.	Organic and inorganic	Removal of suspended solid. Nitrobodies adsorption.

et al. 1976]. TNT production involves three distinct processes, nitration, purification and finishing. Schematically, TNT production is shown in Figure 4.3.

Nitration

Nitration of toluene is performed in a series of batch or continuous reactors, with a mixed acid stream flowing countercurrently to the flow of the organic stream. The oleum fed to the last reactor emerges as the spent acid from the first reactor. The acid stream to the second and third reactors is fortified with 60 percent HNO_3. "Yellow water" is added to the second reactor. Approximately 5 percent [Hudak and Parsons, 1977; Patterson, et al. 1976] of the TNT from the third reactor is β or 2,3,4 and γ or 3,4,6 isomers (the favored product is α or the 2,4,6 isomer).

The gases from the nitrator-separator step contain CO, CO_2, NO, NO_2, N_2O and trinitromethane (TNM). These gases are passed through a fume recovery system, for recovery of NO_x as nitric acid, and are then vented through a scrubber to the atmosphere. Final emissions contain unabsorbed NO_x as well as TNM. Rated capacity for a typical fume recovery system operation for the continuous process is indicated to be 272 Kg HNO_3/hour [Hudak and Parsons, 1977]. Approximately 245 Kg NO_x/Mg TNT are generated by the nitrators of which 9.6 Kg NO_x/Mg TNT are vented to the atmosphere [Hudak and Parsons, 1977]. In addition, unsymmetrical "meta" isomers, as well as oxidation products, are generated. "Red water", the main source of pollution, is the by-product of the treatment of "meta" isomers in the purification step.

The pollutants from TNT production can be alleviated by improving the chemical reaction, to decrease the decomposition of nitric acid, the oxidative side reactions, and the formation of unsymmetrical isomers. For this purpose, a low temperature (-10°C) nitration, to dinitrate toluene, followed by a high temperature (90°C) trinitration is recommended [Hill, et al. 1976; Haas, et al. undated]. The major changes from the existing process would occur in the dinitration step. It has been shown that lowering the temperature from 33°C to -8°C in the dinitrator step results in reduction of "meta" isomers concentration in the

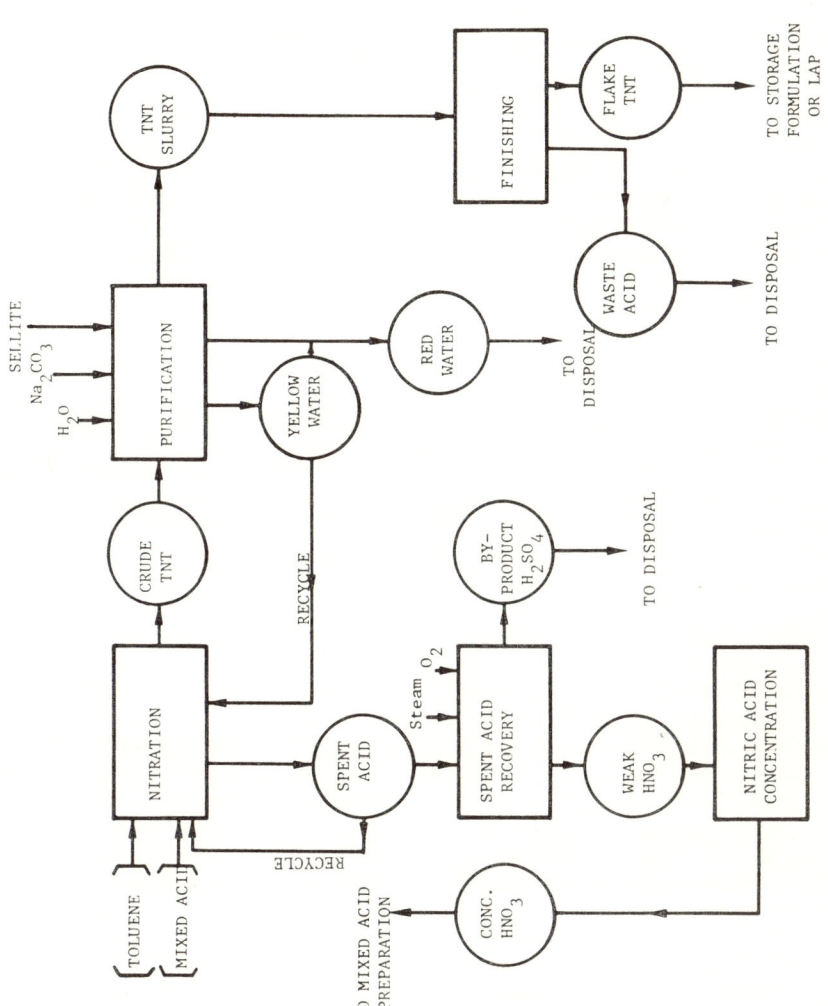

Figure 4.3. Flow Chart for TNT Production (Hudak and Parson, 1977).

product from 2.4 percent to 1.8 percent [Haas, et al. undated].

The fume recovery system for gaseous emissions in the nitration separation step must be designed to operate efficiently at the maximum production level of TNT.

Purification

The crude TNT from the nitration step is washed with water to remove free acids. This step is accomplished by using a countercurrent flow of water. The TNT is then neutralized with soda ash and treated with a 16 percent aqueous sodium sulfite (sellite) solution to remove the nitro group in the meta position, forming highly soluble sodium salts of the corresponding dinitrotoluene sulfonic acid. The TNT is then treated in a series of countercurrent extractors with H_2O and then transferred to the finishing process as a slurry.

The waste streams from the purifications are "yellow water," "red water" and "pink water." "Yellow water" is generated as a result of the first water wash. Some of this acidic effluent is returned to the dinitration step and the rest is combined with other process waste waters for treatment. "Red water," the effluent of sellite treatment and subsequent washing of TNT, consists of 27.6 percent water, 17.3 percent organics, 5.2 percent $NaNO_x$, and 2.9 percent Na_2SO_x [Hudak and Parsons, 1977]. The generation of "red water" amounts to 0.34 kg/kg TNT produced and consists of 0.26 kg process water, 0.06 kg organics (nitrotoluenes and nitrotoluene - sulfonic acid slat), and .02 kg dissovled organics (Na NO_x and $NaSO_x$) [Hudak and Parsons, 1977]. "Pink water" is the waste stream generated from TNT manufacturing as well as LAP operation. "Pink water" arises from the nitration fume scrubber discharge, "red water" concentration distillate, finishing operation hood scrubber and wash down effluent and, possibly, spent acid recovery waste.

Control strategy for reduction of "red water" has been discussed in the nitration section. The traces of TNT in "pink water" and "yellow water" should be recovered. It is possible to purify these streams by treating the waste waters with surfactant, thus forming a precipitate with TNT which can easily be removed by filtration [Okamota, et al. 1979]. The "pink water" resulting from TNT spent acid recovery represents 15-118 kg/day of TNT for a typical plant [Hudak and Parsons, 1977]. In addition "pink water" contains DNT which, by improving the nitration process, should be reduced to a tolerable level.

Finishing

In this process TNT is solidified on a water-cooled flaker drum or belt and is then scraped from the cooling surface with a blade.

The waste stream from the finishing process is mainly waste from spillage, floor drainage, and washings from the finishing area.

The recovery of TNT from these waste streams should be considered if TNT appears in any large quantities in these streams.

Table 4.2 is the summary of the pollution problems associated with TNT production and suggested control alternatives.

Recommended Areas for Pollution Control Research in TNT Production

After evaluating the process technology, the following areas are recommended for further research and development.

- Research on kinetics of the nitration reactions, which will permit alleviation of pollutants by adoption of a low temperature denitration stage.

- Material balances to identify the optimal design for the fume recovery system in the nitration-separation process.

- Recovery of TNT in the "pink water" by application of foam separation to an aqueous solution.

NITROCELLULOSE PRODUCTION

Nitrocellulose (NC) is the second largest product from the military sector of the U.S.

Table 4-2 Pollutants and Control Options in TNT Production

Process	Pollutants	Sources in process	Nature of pollutants	Pollutant control strategy
Nitration	CO, CO_2, NO, N_2O, NO_x TNM unsymmetrical "Meta" isomers of TNT	Nitrator-Separator	Inorganic and organic, gases	Low temperature dinitration stage to reduce "meta" isomers and oxides of nitrogen and carbon.
		Fume recovery system	Inorganic and organic gases	Change the design as to operate at maximum TNT production capacity
Purification	"Yellow water" acidic effluent traces of TNT	First water wash	Inorganic and organic liquid	Removal of TNT from all waste waters by application of foam separation. "Red water" generation will be considerably lower when "meta" isomers formation is decreased.
	"Red water" $NaNO_x$, Na_2SO_x nitrotoluenes and nitrotoluenes sulfuric acid	Sellite wash	Inorganic and Organic liquids	
	"Pink water" DNT, TNT, all pollutants present in red water	Nitration fume scrubber discharge "Red water" concentration distillates. Finishing operation hard scrubber Wash down effluent	Inorganic and organic liquid	
Finishing	Spillage, drainage.			

explosive industry, with a 1969-71 production level of nearly 12,400 tons/month {Patterson, et al. 1976]. NC is the fundamental ingredient used in the production of all gun propellants and many rocket propellants. Cellulose mononitrate, dinitrate and trinitrate have nitrogen contents of 6.75, 11.11, and 14.4 percent respectively, representing progressive replacements of the -OH groups in the cellulose unit by ONO_2 groups. Typically, there are three grades of NC; proxylin, pyrocellulose, and guncotton with 8-12 percent, 12.6 percent, and 13 percent minimum nitrogen content respectively. The preparation of all grades of NC require the same procedure, with the exception of the acid mix used in the nitration process [Mark, 1966]. Manufacturing of NC consists of two processes, nitration and purification. An overall flow chart of NC production is depicted by Figure 4.4.

Nitration

In this process, pre-purified pulp is added to a mixture of nitric acid and sulfuric acid in "dipping pots." Since the nitration reaction is exothermic the vessel is cooled, with the operating temperatures varying from 30 to 40°C. In a continuous process, approximately 68 to 70 kg/min of crude NC is produced.

The nitration is followed by centrifuging the crude NC to remove spent nitrating acids The wrung NC is then dumped into "drowning tubs" filled with water to stop the reaction. The nitrogen content of NC is held between 10.5 percent to 13.8 percent nitrogen [Hudak and Parsons, 1977].

The pollutants emitted from the nitration process are 0.65g SO_x and 1.05g NO_x per kg NC from the reactor pots and 32.5g SO_x and 14.5g NO_x per kg NC from the acid concentrators. The NO from the reactors and centrifuge are vented to an absorber where NO is oxidized and absorbed in the water, generating weak nitric acid. Most of the waste water from the nitration process is due to clean ups. Thus, wastewaters may be expected to have a low pH, relatively high levels of NO , and suspended solids.

Nitration of cellulose in "dipping pots" can be accomplished at 17 to 45°C [Miles, 1955], depending on the nature of the desired product.

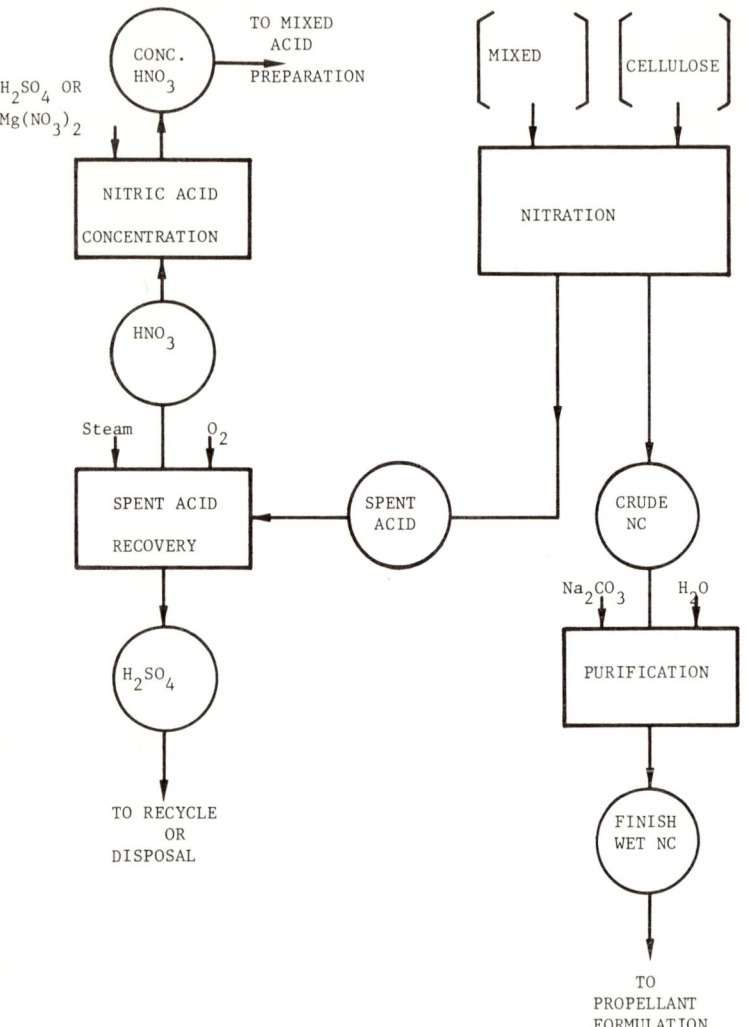

Figure 4.4. Flow Chart for Nitrocellulose Production. (Hudak and Parsons, 1977).

Julian Linares (1966) carried out this nitration at 29°C, producing cellulose nitrate of 13.35 percent N content. Thus, it is evident that the nitration temperatures are flexible, and yield end-products, of comparable quality. The optimal operating temperature must be considered for reduction of SO_x and NO_x emissions. Furthermore, the possibility of converting the waste acid from the nitration process to sulfuric acid must be considered. Such a process involves the reaction of roasting gases containing SO_2 at temperatures exceeding 250 °C with the acid waste (mostly H_2SO_4 and HNO_3) and water spray, to form sulfuric acid [Rensh and Jenthe, 1967]. Absorption of NO_x from the acid concentrator may also prove beneficial.

Purification

The purification of NC involves five processes: boiling tub house, beater house, poacher house, blender house, and final wringer house. In the boiling tub house, unstable sulfate esters and nitrates of partially oxidized cellulose are destroyed by acid hydrolysis. In the beater house, the NC is reduced to a physical state more amenable to purification, by Jordan beaters. In the poacher house, the NC is treated with soda ash, and unpulped fibers are removed. The function of the bleacher house is to sample and regulate the quality of the final product. In the final wringer house, the NC slurry is centrifuged to approximately 30 percent moisture content.

Due to large quantities of process water used during the manufacturing of NC, the treatment and disposal of wastewater is a formidable problem. Presently, overflow from the settling pits flow to the waste acid neutralization facilities, where $CaCO_3$ is added to neutralize the acid. After neutralization, the material is either discharged directly or transferred to settling lagoons. Approximately 13.6×10^6 kg $CaSO_4$ sludge is generated yearly, as a result of waste acid neutralization, at one NC production plant [Hudak and Parsons, 1977].

Wastewaters from the beater, poacher, and blender houses flow to another pit area, where NC fines settle out. Effluent from the pit is either recycled to the wash lines or is discharged. The major portion of total suspended solids in the wastewater discharged is NC fines. One source lists the following NC losses during NC purification [Hudak and Parsons, 1977]:

> Boiling tub house — 68.2 kg/day
> Jordan beater house — 295 kg/day
> Poacher house — 295 kg/day

Nitrocellulose fines lost during the boiling tub, Jordan beater, and poacher house operations constitute inefficiencies in NC purification. NC fines can be recovered from wastewaters by simple filtration, thus reducing the suspended solid content as well as improving the efficiency of the process.

It is, perhaps, not possible to recycle all the water used in the purification of NC but, it may be possible to reduce the volume of the wastewater. For this purpose an in-depth study of NC purification is necessary. Such studies must concentrate on methods of purification with minimum water usage. Thus, an overall material balance of the purification step will be useful.

Table 4.3 is a summary of the pollution problems associated with NC production and suggested control alternatives.

Recommended Areas for Pollution Control Research in Nitrocellulose Production

After evaluating the process technology, the following areas are recommended for further research and development:

- Better understanding of the kinetics of the cellulose nitration will indicate methods of operation to minimize NO_x and SO_x formation.

- Better understanding of absorption of SO_x, and addition of an absorption tower to the acid concentrator to recover NO_x.

- Research on filtration operations and filtration units for effective recovery of NC fines from boiling tub, Jordan beater and poacher houses.

Table 4-3. Pollutants and Control Options in NC Production

Process	Pollutants	Source in process	Nature of pollutant	Pollutant Control Strategy
Nitration	SO_x, NO_x, NO_3 suspended solid	Reactors, nitration vessel.	Inorganic gases and solid.	Improve reaction by temperature adjustment to reduce generation of oxides.
	Waste acid HNO_3, H_2SO_4	Centrifuge	Inorganic acids	Conversion to sulfuric acid by spraying acid waste by water in presence of roasting gases containing SO_2.
Purification	NC fines	Boiling Tub house Jordan beater house poacher house	Organic, suspended solids	Removal by filtration from waste streams.
	Process water, acidic.		Inorganic acids	Reduction of process water.

- Reexamine and modify flow streams to optimize water usage by recycle and other techniques in the purification process.

REFERENCES

Atroshchenko, V., et al. *Simulation of Process and Ammonia Oxidation Reaction*, E.A. *Int. Congr. Chem. Eng. Chem. Equip. Des.*, Autom. 1979.

Dorfman, A.D., et al., *Methods of Ammonia Oxidation Control*, Artom. Khim. Proizvod. (Moscow), 1978.

Haas, R., et al. *Low Temperature Process for TNT Manufacture Part 2 Pilot Plant Development, Industrial and Laboratory Nitration*, ACS-Symposium Series 22, American Chemical Society pp. 272.

Hill, M., et al., *Low Temperature Process for TNT Manufacture Part 1 Laboratory Development, Industrial and Laboratory Nitration*, ACS-SymposiumSeries 22, American Chemical Society, 1976, pp. 253.

Hudak, C. and Parsons, T. *Industrial Process Profiles for Environmental Use: Chapter 12. The Explosives Industry*, EPA 600/277/023L, Feb. 1977.

Julien Linares, *French Patent* 87097 (1966).

Kirk-Othmer Encyclopedia of Chemical Technology, Vol. 8, H.F. Mark, N.Y., Wiley 1966.

Marzo, L. an Marzo, J. Concentrating Nitric Acid by Surpassing an Azeotrope. *Chemical Engineering*. Nov. 3, 1980.

Miles, F., *Cellulose Nitrate*, New York: Interscience Publishers Inc. 1955.

Okamota, et al. *Application of Foam Separation to Aqueous Solutions of TNT. Part II Removal of Organic Explosives With Surfactant*, Order no. AD-A066118, Avail. NTIS. from Gov. Rep. Annoce. Index (U.S.) 1979, 79 (15), 239.

Patterson, J.W., et al. State of the Art: Military Explosives and Propellants Production Industry Vol.III Waste Water Treatment, EPA/600/2-76/213C, Feb.1976.

Rensh, Werner, Jenthe, Horst, Ger (East) Patent 58293 (1967).

Smith, Norman, and Lockyer, John, British Patent 1538198 (1979).

Zhidkov, B., et al. Study of the Kinetics of Ammonia Oxidation in a Platinum Catalyst Kinetics of Decomposition of Nitrogen Oxides, Khim. Tekhnoloy, 1979.

SUPPLEMENTAL REFERENCES

American Defense Preparedness Assn. Waste Treatment in the Military Explosives and Propellants Production Industry. Washington, D.C., Oct., 1975.

Booz-Allen Applied Research, Inc., A Study of Hazardous Waste Materials, Hazardous Effects and Disposal Methods Vol. II, PB 211-466, 1973.

Bradley, R., et al. Classification of Industries Description and Product Lists, SRI Project ECD-3423 94025 for U.S. Environmental Protection Agency, Menlo Park, California, Stanford Research Institute, Dec., 1974.

Dellmeier, Alex; Stecker, Friedrich, German Patent 1,282,529 (1968).

Hadley, W., et al. Potential Pollutants from Petrochemical Processes, Final Report, Contract 68-02-0226, Task 9, MRC-DA-406. Dayton Ohio, Monsanto Research Corp., Dayton Lab., Dec., 1973.

Processes Research, Inc. Air Pollution from Nitration Processes, Contract No. CPA 70-1, Task 22, Cincinnati, Ohio, March 1976.

CHAPTER 5

IRON AND STEEL INDUSTRY

INTRODUCTION

The manufacture of steel involves many processes which require large quantities of raw materials and other resources. Due to the wide variety of products and processes, operations vary from plant to plant. However, the steel industry can be segregated into two major components; raw steel making, and forming and finishing operations. An overview of the iron and steel process is given in Figure 5.1 [U.S. EPA, 1980a].

In the first major process, coal is converted to coke. Nearly all active coke plants are by-product plants which produce, in addition to coke, usable by-products such as coke oven gas, coal tar, crude or refined light oils, ammonium sulfate or anhydrous ammonia, and napthalene. Less than 1 percent of domestic coke is produced in beehive coke making.

The coke from coke making operations is then supplied to the blast furnace process, where molten iron is produced. In the blast furnaces, iron ore, limestone and coke are placed into the top of the furnace and air is blown countercurrently from the bottom. The combustion of coke provides heat which produces metallurgical reactions. The limestone forms a fluid slag which combines with unwanted impurities in the ore. Molten iron from the bottom of the furnace and molten slag, which floats on top of the iron, are periodically withdrawn. The blast furnace flue gas, which has considerable fuel value, is cleaned and then burned in stoves to preheat the incoming air blast to the furnace.

Steel is an alloy of iron, containing less than 1 percent carbon. Steel making consists essentially of oxidizing constituents (particularly carbon) to specified low levels, and then adding

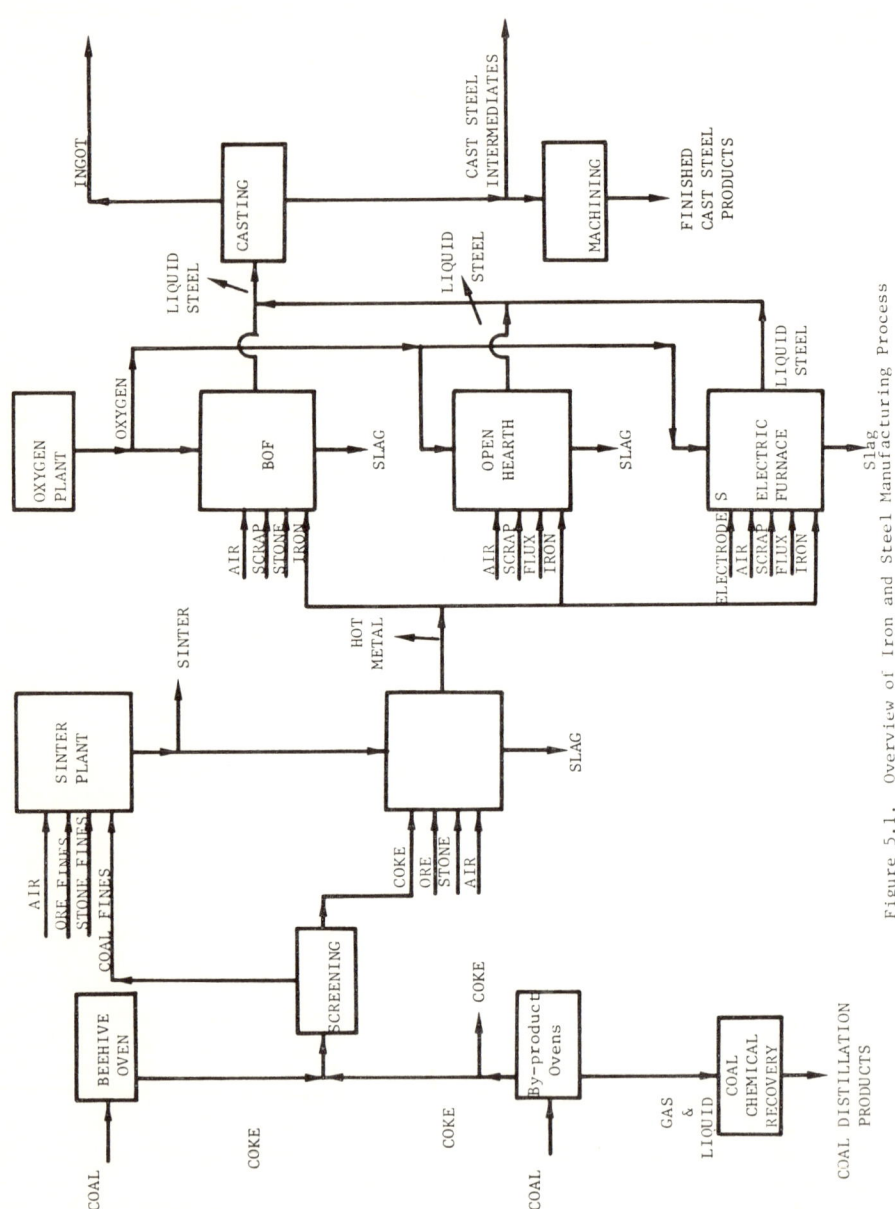

Figure 5.1. Overview of Iron and Steel Manufacturing Process

various alloying elements according to the grade of steel to be produced. The basic raw materials for steel making are hot metal pig iron, steel scrap, limestone, burned lime, dolomite, fluorspar, iron ores, and iron-bearing materials such as pellets or mill scale. The principal steel making processes in use today are the Basic Oxygen Furnace (BOF), the Open Hearth Furnace (OHF), and the Electric Arc Furnace (EAF).

Hot forming (including continuous casting) and cold finishing operations follow the steel making process. These operations are so varied that simple classification and description is difficult. In general, hot forming primary mills reduce ingots to slabs or blooms, and secondary hot forming mills reduce slabs or blooms to billets, plates, etc. Steel finishing involves a number of operations that basically impart desirable surface or mechanical properties to the steel. Correct surface preparation is the most important requirement for satisfactory application of protective coatings to the steel surface. The steel surface must be cleaned at various production stages, to insure that the oxides which form on the surface are not worked into the finished product. The pickling process chemically removes oxides and scale from the surface of the steel by the action of inorganic acids. This method is the most widely used, due to comparatively low operating costs and ease of operation. Pickling will be the only finishing operation considered here.

COKE MAKING

Figure 5.2 provides a flow diagram of a coke plant, the processes required including:

1. Coal mining and transportation
2. Coal preparation
3. Charging of coal
4. Coking
5. Pushing and quenching
6. Coke handling and tar condensation

Coal dust is emitted during size reduction and transportation of the coal. Emissions are estimated to be 5 kg/ton of coal. Particles are generally less than 0.1 mm in size, with suspended fractions in the range of 1 to 10 μ [Barnes, et al. 1970]. The wastewaters produced from coal

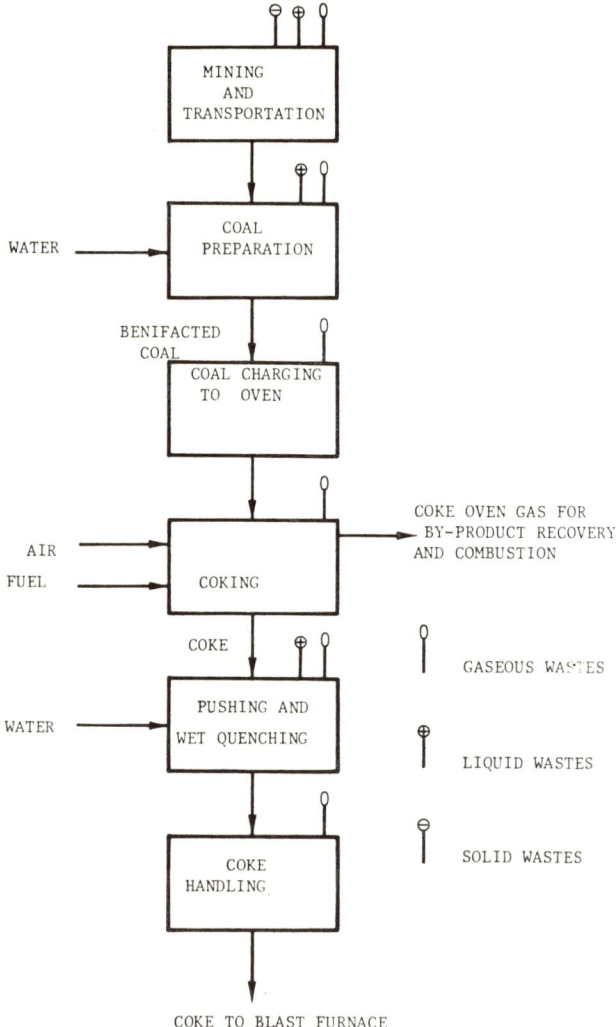

Figure 5.2. Coke Making

preparation may contain up to 200 g/l of suspended particles (28 mesh to colloidal dimensions) and most can be eliminated by thickeners, cyclones, and filters [Keystone Coal Industry Manual, 1974]. Trace elements contained in the wastewater include arsenic, cadmium and lead.

Coke is the residue from the destructive distillation of coal. Coal is heated in the coke ovens (with no air in the oven) by adjacent chambers or flues using; 1) some of the gas recovered from coking operations, 2) cleaned blast furnace gas, or, 3) a mixture of coke oven and blast furnace gases. During carbonization, about 20 to 35 percent by weight of the initial coal charge is evolved as mixed gases and vapors, which are vented through an opening at the top of the oven. From one ton of coke, about 544-634 kg blast furnace coke, 45-91 kg coke breeze, 270-325 m^3 coke gas, 30-45 kg tar, 9-13 kg $(NH_4)_2SO_4$, 54-312 kg ammonia liquor and 9-15 kg light oil are produced [Keystone Coal Industry Manual, 1974]. Potentially hazardous emissions from the coke plant include CO, amines, organometallics, tar and soot, carbonyls, hydrogen cyanide, sulfur compounds, etc. [Cavanaugh, et al. 1974]. The use of refined or cleaned coal as a raw material and better knowledge of coking reactions can reduce or eliminate some of these pollutants.

Most of the particulates generated in hot coke quenching operations are collected and reused within the plant, and do not constitute a significant pollution problem. However, wastewaters from wet coke quenching operations contain high concentrations of ammonia, oil and grease, and phenol (all of the three exert a biochemical oxygen demand), plus cyanide, sulfide and suspended solids. Table 5.1 gives an analysis of quench wastewaters samples [U.S. EPA, 1977].

Some the treatment technologies used for control of pollution from the quench waters are [U.S. EPA, 1974]:

1. Distillation with ammonia recovery of waste ammonia liquor
2. Alkaline ammonia stripping
3. Neutralization
4. Settling
5. Air flotation

6. Clarification, with vacuum filtration of sludge

7. Filtration

8. Carbon adsorption

Table 5.1 Average Analysis of Quench Water Samples

Contaminant	Concentration (ppm)
Phenols	776
Sulfates	1,066
Chlorides	1,954
Total Ammonia	2,517
Cyanides	98
Total Solids	5,214

Conversion of hot coke quenching from the wet to the dry process would: 1) eliminate air and wastewater emissions from the wet quenching process, and 2) provide additional potential for particulate emission control, since control of these emissions is part of the dry quenching design [U.S. EPA, 1976]. Pollution control costs are not significant for the dry process. The hot quench gases can be cooled, recovering useful energy. Capital and operating costs are significantly higher than wet quenching operations, by up to $10.70 per ton of coke. Dry coke quenching is claimed by Russian authors to produce a higher-grade coke, thus reducing the coke demand in the blast furnace. However, this claim needs to be demonstrated for U.S. coals.

A detailed description of the dry coke quenching process has been given [EPA, 1976]. In dry quenching of coke, the hot coke pushed from the ovens is cooled in a closed system. Dry quenching uses "inert" gases to extract heat from the incandescent coke by direct contact. The heat is then recovered in waste heat boilers or by other techniques. The "inert" gases can be generated from an initial intake of air which reacts with the to coke to form a quenching gas containing 14.5% CO_2, 0.4% O_2, 10.6% CO, 2% H_2 and 72.5% N_2. After cooling, the gases pass through two dust recovery cyclones (where particulates are collected at ~400-600 lb/hr) before being recycled. The particulates consist mainly of carbon dust, which is burnt as solid fuel. Part of the gases after particulate removal may be sent for removal of evaporated cyanides and SO_x, to prevent accumulation. Additional research is needed to study the applicability of dry quenching of U.S. coals and the cleaning of recycle gases using systems similar to those used in by-product coke recovery operations.

COKE BY-PRODUCT RECOVERY

Figure 5.3 shows the various coke by-products recovery operations [U.S. EPA, 1977]. The condensate from the cooled coke oven gases contains tar and ammonia liquor, which are decanted. The uncondensed gases are reheated and passed on to the ammonia scrubber. Some of the tar derivatives recovered include creosote oil, creosols, naphthalene, phenol and medium and hard pitch. Phenol is recovered from the weak ammonia liquor by scrubbing with benzene or light oil. The phenolized benzene, or light oil, is contacted with caustic soda to extract sodium phenolate. The sodium phenolate is neutralized with CO_2 to liberate crude phenols.

The weak ammonia liquor is heated to remove "free" ammonia and is then contacted with lime solution to remove and strip NH_3 as vapor. These vapors are sent to the ammonia absorption process. The waste liquor generated in the ammonia still contains sulfide as H_2S. These wastes may contain up to 20 percent by volume of $Ca(OH)_2$ [U.S. EPA, 1977]. It is suggested that $Ca(OH)_2$ be recovered from the waste liquids and recycled to the ammonia still.

Ammonia vapors from the primary cooling and heating operations and the ammonia still are absorbed in a dilute H_2SO_4 solution. The $(NH_4)_2SO_4$ product is crystallized from the resulting slurry. Gases, leaving the ammonia absorber, are scrubbed with recycled wash oil to remove light oil. The light oil is refined to produce benzene, toluene and xylene. The uncondensed "coke oven gases" are sent to a gas holder and contain mainly CO_2, CO, N_2,

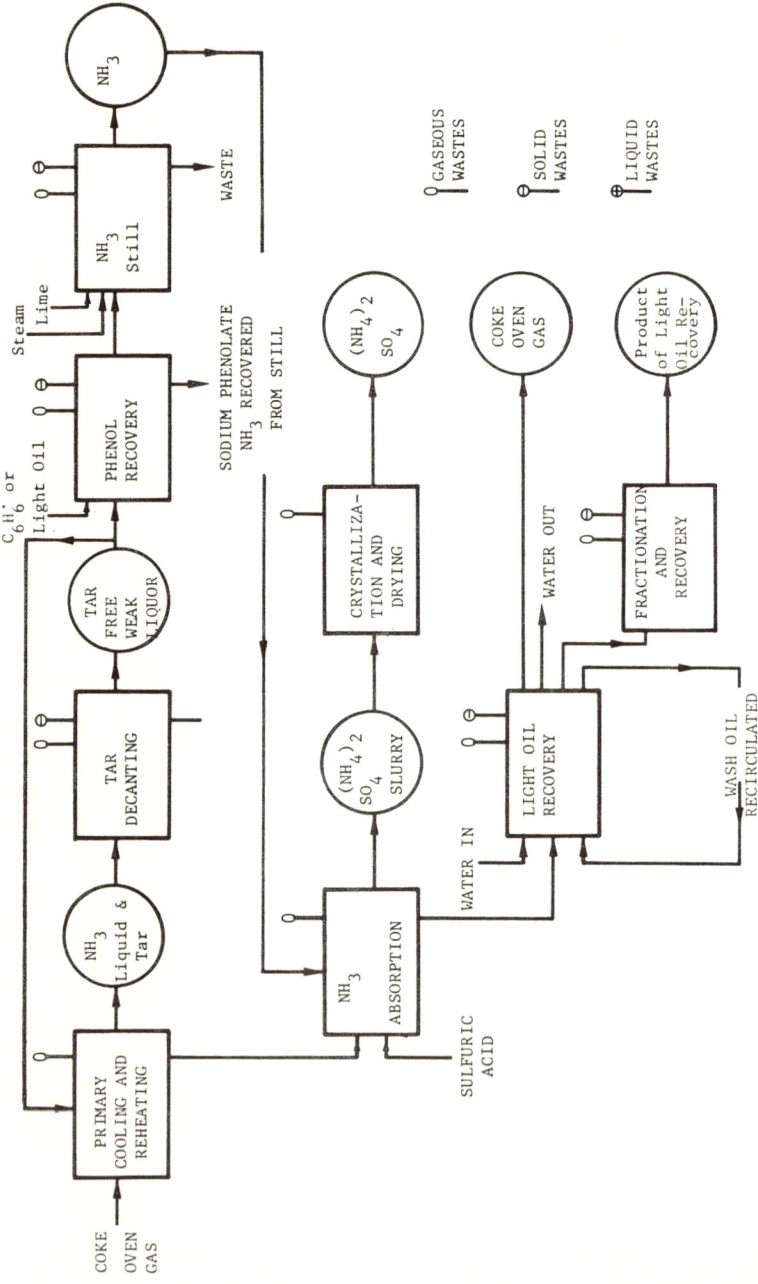

FIGURE 5.3. Coke By-Products Recovery (U.S. EPA, Feb. 1977).

H_2, O_2 and H_2S. The composition is given in Table 5.2. It is suggested that H_2S be removed prior to ammonia absorption, using the Claus or Stretford process [Dunlap, et al. 1973].

Table 5.2 Coke Oven Gases (After By-Product Recovery) [Ess, 1948; Wilson, et al. 1980]

Component	Kg/ton of Coal
CO_2	10.42
CO	31.54
H_2	13.66
N_2	3.85
O_2	7.17
H_2S	3.25

One of the major pollutants from light oil recovery is cyanide [Allen, 1979]. HCN contained in the gas leaving the cooling tower at one site amount to 0.56 lb/ton of coke. It is suggested that cyanide by removed by adsorption or other suitable methods. About the same amount of cyanide is expected to be in the cooler wastewaters. Biological systems followed by nutrient addition in the form of phosphoric acid have been reported to produce a reduction of cyanide of up to 95 percent [Hofstein and Kohlmann, 1980].

IRON MAKING

Blast furnaces are large cylindrical structures in which molten iron is produced by the reduction of the iron bearing ores with coke and limestone. Reduction is promoted by blowing heated air into the lower part of the furnace. As the raw materials melt and decrease in volume, the entire mass of the furnace charge descends. Additional raw materials are added (charged) at the top of the furnace, to keep the amount of raw materials within the furnace at a constant level.

Iron oxides react with the hot CO from the burning coke, and the limestone reacts with impurities in the iron ore and coke to form molten slag. These reactions start at the top of the furnace and proceed to completion as the charge passes to the bottom of the furnace. The molten slag, which floats on top of the molten iron, is drawn off (tapped) by way of a "tapping hole." Blast furnace operations within the U.S. produce greater than 99 percent of the basic iron. The total rated capacity of all U.S. plants is 321,847 tons/day.

The gases which are produced in the furnace are exhausted through the top of the furnace. These gases are cleaned, cooled, and then burned to preheat the incoming air to the furnace. Generally, gas cleaning involves the removal of the larger particulates by a dry dust collector, followed by a variety of wet scrubbers for finer particle removal. Many of the same pollutants found in coke plant wastewaters are also found in iron-making wastewaters. The phenolic pollutants found in iron-making wastewaters are attributable to the coke used in the iron-making process. Cyanide and ammonia (reaction products formed within the furnace or transferred from the coke charge to the furnace gases) are carried over with the gas stream, and transferred to scrubbing waters. Table 5.3 lists some of the pollutants in these scrubbing waters [U.S. EPA, 1980b].

Table 5.3   Analysis of Blast Furnace Scrubber Wastewaters

Pollutant	Concentration mg/l
Ammonia	63.2
Cyanide	16.9
Phenols	2.77
Fluoride	23
Suspended Solids	693
pH	7.1 - 8.3

The cleaned gases are then cooled with direct contact sprays in large gas cooling vessels.

About 90 percent of the existing blast furnace wastewater treatment systems include recycle (after thickening) and discharge only about 5 to 10 percent of the process flow. The dewatered solids from thickener are either sent to the sintering operations or to off-site disposal. These solids contain heavy metals. The effluent concentrations are listed in Table 5.4.

Table 5.4 Wastewater Thickener Underflow [U.S. EPA, 1980b]

Pollutant	Concentration mg/l
Cu	0.19
Pb	2.11
Ni	0.10
Ag	0.023
Zn	27.5

It is suggested that the gas cleaning system be modified to use few and, if possible, no wet scrubbers. For fine particle collection the possibility of using electrostatic filters or magnetic filters should be studied. The cleaned gases are generally cooled in spray towers (before being burnt) thus producing more wastewaters. Instead, it is suggested that 1) cleaned gases be burnt directly, or 2) be cooled by heat exhcange with air feed to the blast furnace and then burnt.

It was suggested for the coke by-product recovery process that organic compounds be removed by adsorption on activated carbon. Similar systems are proposed for removal of organics from the blast furnace wastewaters. The use of biological oxidation and alkaline chlorination systems for treatment of these wastewaters to remove HCN is also suggested.

Hofstein and Kohlman (1979) report that one of the non-U.S. plants they visited uses Caros Acid (per-monosulfuric acid - H_2SO_5) to reduce cyanide and phenol levels in the blast furnace wastewaters to 0.2 mg/l HCN and 0.5 mg/l phenol. The normal influent level of phenol at that plant was 2 mg/l. The addition of polyphosphate in the cooling towers had been reported to aid in CN removal, but at CN levels above 10 mg/l it was not effective. Another plant theorized, based on operating experience, that the formation of metallo-cyanide complexes adsorbed in the sludge reduced cyanide levels from 0.2 mg/l to 0.1 mg/l. Another plant, in West Germany, uses aeration prior to discharge to the clarifiers, as an integral part of the gas cleaning water recirculation system. The purpose of aeration is to strip CO and CO_2 from the water, and to precipitate $CaCO_3$. A portion of the clarifier sludge is recycled, to act as a seed and enhance precipitation and sedimentation.

Osantowski and Gienopolos (1979) obtained pilot plant data on the applicability of advanced waste treatment methods for upgrading blast furnace wastewaters. The treatment methods investigated included: alkaline chlorination, clarification, filtration (dual media and magnetic), ozonation, activated carbon and reverse osmosis. Alkaline chlorination consisted of elevating the pH of the oncoming wastewaters to 11.0-11.5 with addition of sodium hypochlorite for oxidation of cyanide; followed by neutralization for ammonia removal. The settled effluent was dechlorinated by activated carbon. These studies showed that above a $Cl_2:NH_3$ ratio of 7.3:1, NH_3 levels dropped sharply to as low as 0.48 mg/L, NH_3 removal was high and all other oxidizable contaminants were also removed. Further research is needed to optimize the $Cl_2:NH_3$ ratio to obtain the best ammonia removal. Ozonation was determined to be effective in meeting discharge levels. Prefiltered wastewater was elevated to pH 10.5 prior to ozonation. Th ozone dosage varied from 0-1,500 mg/l of wastewater treated, with contact times from 60-240 minutes. NH_3 and cyanide levels of 3.1 and 0.001 mg/l were achieved. Research is required to establish the optimal pH and contact time.

Metals in the raw wastewaters may be best removed by sulfide precipitation. The use of excess sulfide in a treated effluent may cause an objectionable odor problem, however. A decrease in pH might affect personnel if the wastewater become acidic, and evolved H_2S. In veiw of these objections, a ferrous sulfide slurry may be a good choice since it does not dissassociate readily, thus, controlling the presence of excess sulfide. To this date, there are no furnace wastewaters systems currently using sulfide precipitation [U.S. EPA, 1980b], even though this technology has been demonstrated in the metal finishing industry. It is suggested that the applicability of sulfide precipitation to remove metals from blast furnace wastewaters by investigated.

There are two types of dust materials collected from the blast furnace operations; 1) dry-collected dusts are obtained from cyclone dust collectors, and 2) wet-collected dusts are obtained from wet scrubbers after reducing the water content by thickeners and filters. The bulk of this material is used as landfill, even though some is utilized in the sintering operations. Coke, pellet, and BOF slag fragments are the predominant components of the dry dust materials and include significant quantities of hematite, magnetite, graphite, calcium carbonate, wusite and silica. The size distribution of these dusts ranges from 2.5 to 0.014 mm. Wet-collected dusts are similar to dry-collected dusts and average 24 percent by weight Fe and 45 percent by weight C, with about 68 percent of the material less than 60 μm in size. The high and variable carbon content of this material makes it difficult to use in the sintering operations. Furthermore, the high amounts of zinc in the wet-collected dusts can impede the blast furnace operations, if recycled. A briquetting method for dezincing these dusts has been developed [Allen, 1979]. Typical analysis of salable zinc oxide dust is 63.3 percent Zn, 6.9 percent Pb, and 1.2 percent Fe.

Another possible, alternative dezincing process may involve extracting the zinc from collected dusts. The Waelz process is a direct reduction technique that has been used for 20 years to refine low grade zinc ores. The Berzelium and Lurgi companies conducted a large scale experiment in Dinsburg, Germany, that used iron and steel material dusts [Rausch and Serbert, 1948]. They observed that dezincification was possible with a continuous process, with 95 percent of the zinc and 50 percent of the alkalis removed and 95 percent of iron metalized. This process was found to operate more economically on a lower throughput (10^5 tons/year) than other direct reduction processes [U.S. EPA, 1979] and, it is less sophisticated since it does not require pelletizing before reduction.

Advances in technologies such as auxiliary fuel injection, higher hot blast temperatures, moisture injection, and oxygen enrichment have been major factors in increasing the efficiency of the blast furnace process. Based on a statistical analysis technique of certain operating parameters, Quigley and Sayles (1974) concluded that the coke stability is improved by increasing bulk density and coking time, and improving coal grinding. They also found that an increase of 1.5 percent in ash content of the coke resulted in over 50 percent off-quality iron over a two-day period. Thus studies aimed at improving blast furnace performance by improving the coking process appear appropriate.

Slags from the blast furnace usually come into contact with water immediately after their removal from the blast furnace. The reaction between slag and water produces H_2S and SO_2, among other gases, which even though small in quantity, can cause odor problems. Knowledge of the mechanisms of H_2S and SO_2 formation is limited. It is believed that SO results primarily from the oxidation of H_2S. It is desirable to suppress the formation of sulfurous gases, primarily H_2S. In laboratory experiments, Kaplan and Rengstroff (1973) showed that treatment of the slag with the oxidizing materials Fe_2O_3, CO_2, $CaCO_3$, decreased the emission of H_2S. Inplant trials, where commercial batches of blast furnace slag were treaed with steam prior to water quenching the hot solid slag and molten slag, the steam treatment resulted in an increase in the emission of H_2S.

One important process modification would be to optizime slag quenching methods to decrease H_2S emissions. This would require a better understanding of H_2S and SO_2 formation mechanisms ions during slag quenching, and the effect of various additives.

Another option is to produce elemental sulfur from H_2S by the Claus or Stretford Process. If HCN is found to cause fouling of the Claus catalyst beds or decrease the life of Stretford absorption solution, HCN may have to be removed from the feed gases by the polysulfide process, which is known to have removal efficiencies of up to 90 percent of HCN [Hill, 1945].

Cyanides are thought to form in the blast furnace according to the following reactions:

$$K_2O + 3C + N_2 \rightleftarrows 2KCN + CO \tag{5.1}$$

This reaction is favored at high temperatures and the absence of CO. The oxidation of cyanides takes place according to the following reaction:

$$2KCN + 4CO_2 \rightleftarrows K_2CO_3 + N_2 + 5CO \tag{5.2}$$

It would appear that conditions favorable for the formation of cyanides, i.e., the coexistence of coke, alkali oxides, and N_2 at high temperatures and in a reducing atmosphere, are inherent in the operation of the iron blast furnace. In contrast, it may be possible, at least in principle, to modify the conditions in the stack region, so that oxidation of cyanides is promoted. Very little work has been done, however, on the kinetics of the reactions in the formation and oxidation of cyanides in the blast furnace. Sohn and Szekely (1943) assumed mass transfer to the KCN particle to be rate controlling, and concluded that there may be a real possiblity of reducing the net emission f cyanides from the blast furnace by the appropriate manipulation of the temperature and CO/CO_2 profiles within the stack. It is suggested that efforts be directed toward obtaining kinetic information on the formation and oxidation of cyanides in the blast furnace, and an optimizing stack design and operation to reduce the total cyanide in the effluent gases.

STEEL MAKING

Steel is an alloy of iron containing less than 1.0 percent carbon. Steel making is basically a process in which carbon, silicon, phosphorous, manganese, and other impurities present in the raw hot metal or steel scrap, are oxidized to specific minimum levels. The hot steel is then either teemed into ingots or transferred to a continuous casting or pressure casting operation for direct conversion into a semi-finished product.

The basic raw materials for the steel making processes are hot metal or pig iron, steel scrap, limestone, burned lime, dolomite, fluorspar, iron ores and iron bearing minerals such as pellets, mill scale and waste solids from the furnaces.

86 PROCESS MODIFICATIONS

Various types of steel are manufactured by adding alloying agents either to the hot charge in the furnace or to the ladle of steel after the hot steel is "tapped" from the furnace. The three types of steel making processes are:

1. Basic Oxygen Furnace (BOF)
2. Open Hearth Furnace (OHF)
3. Electric Arc Furnace (EAF)

A comprehensive discussion of the various operations that constitute each of these processes is given by McGannon (1964).

The large quantities of air borne gases, dusts, smoke, and iron oxide fumes generated in the steel making processes are collected and contained by various gas cleaning systems. Depending on the type of gas cleaning system used, wastewater discharges or sludge generation can result. The basic gas treatment systems used are:

<u>BOF</u> Semi-wet
Wet-suppressed combustion
Wet-open combustion

<u>OHF</u> Semi-wet
Wet

<u>EAF</u> Semi-wet
Wet

The semi-wet air pollution control systems use water to partially cool and condition the waste gases and fumes, prior to final particulate removal in dry collectors such as precipitators or baghouses. Wet air pollution control systems use water not only to cool the waste gases, but also to scrub the fume particles from the waste gases. The BOF uses combustion systems, suppressed or open, to control CO emissions.

Two variations of the BOF process are the Kaldo process and the Q-BOP furnace. At present, there is only one Kaldo installation and only three Q-BOF installations in the United States. The waste products from the BOF steel making process include airborne fluxes, slag, carbon monoxide and dioxide, and oxides of iron (FeO, Fe_2O_3, Fe_3O_4) emitted as submicron dust. Also, when hot metal (iron) is poured into ladles or the furnace, submicron iron xide fumes are released and some of the carbon in the iron is precipitated as graphite, commonly called "kish." Approximately 1 to 2 percent of the ingot steel is lost as dust. The primary gas constituent emitted from the BOF during the oxygen blowing cycle is CO. CO will burn outside of the BOF, if allowed to come in contact with outside air (open combustion). If outside air is prevented from coming in contact with the CO gas combustion is retarded (suppressed combustion).

In the Open Hearth Furnace process, steel is produced in a shallow rectangular refractory basin, or hearth, enclosed by refractory lined walls and roof. OH furnaces can use an all scrap steel charge; however, a 50 percent scrap/50 percent hot metal charge is typical. Fuel oil, coke oven gas, natural gas, etc. are burned as fuel and the hot gases are circulated above the raw material charge. The waste gases are water cooled, scrubbed with recycled water and sent through a electrostatic precipitator before being vented. The waste products resulting from the OHF process are slag, iron oxides as submicron dust, waste gases (consisting of air, CO_2 and water vapor), SO_x and NO_x (due to the nature of certain fuels being burned), and oxides of zinc.

The Electric Arc Furnace (EAF) steel making process produces high quality alloy steels in refractory lined cylindrical furnaces using a cold steel scrap charge and fluxes. The waste products from the EAF process are smoke, slag, CO, CO_2, metal oxides (mainly iron) emitted as submicron particles, and zinc oxides.

Raw wastewater characteristics are given in Table 5.5 [U.S. EPA, 1980b].

The treatment methods presently used do not attempt to remove these toxic metals exclusively [U.S. EPA, 1980b]. Sulfide precipitation to enhance the precipitation of metals from wastewaters, before recycle, is one suggested pollution control approach. There is a need to develop extraction and ion exchange systems to remove toxic metals formed in the steel furnace wastewaters. Suspended solids in steel making wastewaters are very fine, red in color, and principally iron oxide. Magnetic separation can reduce the level of suspended solids to 25-30 mg/l [Centi, 1973].

Table 5.5 Raw Wastewaters Concentrations From Steel Making Operaton (mg/l)

I. **Basic Oxygen Furnace**

    a. Semi-wet (Flowrate - 429 gal/ton)

Suspended solids	345.0
Cu	0.004
Pb	1.5
Zn	1.0

    b. Wet-Suppressed Combustion (Flowrate = 982 gal/ton)

Suspended solids	1500.0
Cr	0.5
Cu	0.25
Pb	15.0
Ni	0.5
Zn	5.0

    c. Wet-Open Combustion (Flowrate = 446 gal/ton)

Suspended Solids	4200.0
Cd	0.5
Cr	5.0
Pb	1.0
Zn	5.0

II. **Open Hearth Furnace**

    a. Semi-wet (Flowrate = 1163 gal/ton)

Suspended Solids	500.0
Cr	0.8
Cu	0.8
Cyanide	0.04
Zn	0.5

    b. Wet (Flowrate = 554 gal/ton)

Suspended Solids	1100.0
Cu	2.0
Pb	0.6
Zn	200.0

Table 5.5 Continued

III. <u>Electric Arc Furnace</u>

    a.   Semi-wet (Flowrate = 60.5 gal/ton)

Suspended Solids	2200.0
Cu	2.0
Pb	30.0
Zn	125.0

    b.   Wet (Flowrate = 3060 gal/ton)

Suspended Solids	3400.0
Arsenic	2.0
Cd	4.0
Cr	5.0
Cu	2.0
Pb	30.0
Zn	125.0

ACID PICKLING

Acid pickling is the steel finishing process in which steel products are immersed in heated concentrated acid solutions to remove surface scale. Based on the type of pickling acid used, this process can be subdivided into:

1. Sulfuric acid pickling
2. Hydrochloric acid pickling
3. Combination acid pickling

Wastewaters are generated by three sources in the pickling process. The largest source is the rinsewater used to clean the acid solution from the product after it has been immersed in the pickling solution. The second source is the spent pickling acid liquor. The spent pickle liquor is a small volume waste, containing high concentrations of acid, iron, and toxic metal pollutants. Wastewater from the wet acid fume scrubbers is the third source.

Rinse water discharge flows can be minimized with cascade or countercurrent rinse systems. These systems reduce water flow, concentrate the pollutants in the last rinsing chamber, and achieve more thorough rinsing [U.S. EPA, 1980c].

The most common method of recovering spent sulfuric acid is acid recovery, by removing ferrous sulfate through crystallization, and rinsing the concentrated acid [U.S. EPA, 1980c]. The spent liquor from hydrochloric acid pickling contains free hydrochloric acid, ferrous chloride, and water. This liquor is heated to 1,050°C and, at this temperature, the water is completely evaporated and $FeCl_2$ decomposes completely into Fe_2O_3 and HCl gas. The iron oxide is separated and removed, and HCl gas is reabsorbed in the water.

Recycle systems, with recycle rates as high as 90-95 percent of the total wastewater flow, are used to control pollutants generated during the scrubbing of acid fumes. Recycle rates are limited by the build-up of solids. Lime and sulfide precipitation are suggested to improve precipitation of metals in the wastewaters.

Temperature and agitation are important operating factors in the pickling process. The temperature of pickling dramatically affects the reaction rate. Agitation is probably the most ignored aid to good pickling. The speed of pickling can be increased significantly by properly agitating the acid bath or the steel product during the pickling operation. One method of agitation is an air-operated, mechanical agitation system. An added benefit of this system is that the evaporation caused by air agitation concentrates, rather than dilutes, the acid bath. It is suggested that process improvements be aimed at decreasing pickling time, and maintaining good agitation and high enough acid concentration levels during pickling.

Table 5.6 lists the various process modification suggested for control of pollution from the iron and steel industry.

RECOMMENDED AREAS FOR PROCESS MODIFICATIONS RESEARCH

After careful evaluation of the process technology, the following areas are recommended for further research and development:

Table 5-6. Pollutants and Suggested Control Strategy Iron and Steel Industry

Process	Pollutant	Source in Process	Nature of Pollutant	Pollutant Control Strategy
Coke Making	Organometallics, Carbonyls, HCN, SO_x, H_2S	Emissions from coke ovens	Toxic gases	Use of refined coal, better control of coking operations.
	Phenols, Sulfates, Chlorides, Ammonia, Cyanides	Wastewaters from coke quenching	Dissolved and suspended solids	Conversion to dry coke quenching with particulate removal
Coke By-Product Recovery	H_2S	Coke oven gases having by-product recovery	----------	Removal of H_2S by Claus or Stretford Process
	HCN	Cooler Wastewaters	----------	Removal of HCN by biological oxidation systems
Iron Making	Metals (Zn, Pb, etc.)	Blast Furnace wastewaters	Suspended solids	Conversion from wet to dry gas cleaning systems - electrostatic, magnetic filters

Table 5-6. Pollutants and Suggested Control Strategy for Iron and Steel Industry (Cont'd)

Process	Pollutant	Source in Process	Nature of Pollutant	Pollutant Control Strategy
Steel making	Zn, Cr, Pb, Cu, iron oxide	Wastewaters from cooling and conditioning of gases from steel making furnaces	Suspended, dissolved solids	Sulfide precipitation of toxic metals & recycle of treated wastewaters magnetic separation of suspended iron oxides, solvent extraction of metals
Acid Pickling	Spent Pickle Liquor (SPL)	----------	Inorganic acids, suspended toxic metals, iron	Improved pickling process by temperature control and good agitation to decrease SPL, recovery of acid
	Rinse Waters	Cleaning of treated product after pickling	Inorganic acids, suspended toxic metals, iron	Minimizing rinsewater discharges by using cascade or counter-current rinse systems

- Use of refined or cleaned coal in coke making.

- Study of the formation of cyanides and carbonyls during coking, in order to prevent or reduce pollutant formation.

- Dry coke quenching studies of U.S. coals, to eliminate quench water contamination.

- The effect of process variables such as coal grinding and coking time on coke produced. This would be used to improve the coke stability, in order to enhance blast furnace performance.

- Evaluation of biological oxidation systems to remove HCN from by-product coke making wastewaters.

- Absorption of H_2S from coke oven gases by Claus or Stretford process, for sulfur recovery.

- Development of better CN removal systems from blast furnace wastewaters by using additives such as Caros acid (H_2SO_5) and polyphosphate, and by aeration.

- Optimization of pH and contact time to improve alkaline chlorination and ozonation to upgrade blast furnace wastewaters.

- Solvent extraction of zinc, or dezincification by Walez process, to recover zinc from blast furnace dusts. Such zinc removal can also facilitate the use of treated dusts in the sintering process.

- Study of kinetics of H_2S and SO_2 formation during blast furnace slag quenching, to develop methods by which H S formation can be decreased.

94   PROCESS MODIFICATIONS

- Study the kinetics of formation and oxidation of cyanides in the blast furnace, to optimize the design and operation of the stack in order to oxidize the cyanide.

- Development of solvent extraction and sulfide precipitation methods to remove toxic metals from the recycled steel plant wastewaters.

- Study the effect of temperature and agitation in order to optimize pickling operations and extend the pickle liquor life.

REFERENCES

Allen, E.J. International Mineral Recovery Ltd. Dezincing Process, Proc. of the First Symp. on Iron and Steel Pollution Abatement Technology, Chicago (1979), EPA-600/9-80-12, PB80-146258.

Allen, G.C., Jr. Environmental Assessment of Coke By-Product Recovery Plant, Proc. of the First Symp. on Iron and Steel Pollution Abatement Technology, Chicago, (1979), EPA-600/9-80-012, PB80-176258.

Barnes, T.M., et al. Evaluation of Process Alternatives to Improve Control of Air Pollution from Production of Coke, Battelle Memorial Institute, Columbus, OH, Jan. 1970.

Cavanaugh, et al. Potentially Hazardous Emissions from the Extraction and Processing of Coal and Oil, EPA-650/2-75-038, April, 1974.

Centi, T.J. A Survey of Wastewater Treatment Techniques for Steel Mill Effluents, in "The Steel Industry and The Environment", ed. by Szekely, J., Marcel Dekker, Inc., New York, (1973).

Dunlap, R.W., et al. Desulfurization of Coke Oven Gas: Technology, Economics and Regulation Activity, in "The Steel Industry and The Environment", Szekely, J., (ed.); Marcel Dekker, Inc., New York, (1973).

Ess, T.J. The Modern Coke Plant, Iron and Steel Engineer, C3-C36, Jan. 1948.

Hill, W. Recovery of Ammonia, Cyanogen, Pyridine, and Other Nitrogenous Compounds from Industrial Gases, in "Chemistry of Coal Utilization, Vol. II", ed. by, Lowry, H.H.; Wiley, New York, (1945).

Hofstein, H. and Kohlmann, H.J. Study of Non-U.S. Wastewater Treatment Technology at Blast Funace and Coke Plants, Proc. of the First Symp. on Iron and Steel Pollution Abatement Technology, Chicago, (1979), EPA-600/9-80-012, PB80-176258.

Kaplan, R.S. and Rengstroff, G.W.P. Emission of Sulfurous Gases from Blast Furnace Slags, in "The Steel Industry and the Environment, ed. by Szekely,J.; Marcel Dekker, Inc., New York, (1973).

Keystone Coal Industry Manual, Mining Information Services of the McGraw Hill Publishing, 1974.

McGannon, E.H., The Making, Shaping, and Treating of Steel, United States Steel Corporation, Eighth ed., 1964), Pittsburgh, Pa.

Osantowski, R. and Geinopolos, A. Physical-Chemical Treatment of Steel Plant Wastewaters Using Mobile Plant Units, Proc. of the First Symp. on Iron and Steel Pollution Abatement Technology, Chicago, (1979), EPA-600/9-80-012, PB80-176258.

Quigley, J.J. and Sayles, N. An Analysis of the Effect of Coke Characteristics and Other Operating Variables Upon Blast Furnace Performance, 33rd Iron Making Conference Proc., Vol. 33, Atlantic City Meeting, April 28 - May 1, (1974).

Rausch, H. and Serbert, H. Benefaction of Steel Plant Waste Oxides by Rotary Kiln Processes, Paper Presented at Sixth Mineral Waste Utilization Symp., Chicago, May 2-3, (1948).

Sohn, H.Y. and Szekely, J. <u>On the Oxidation of Cyanides in the Stack Region of the Blast Furnace</u>, ed. by, Szekely, J., Marcel Dekker, Inc., New York, (1943).

U.S. EPA. <u>Development Document for Effluent Limitations Guidelines and Standards for the Iron and Steel Manufacturing Proposed, Point Source Category, Vol. I</u>, EPA-440/1-80/024-b, December 1980a.

U.S. EPA. <u>Development Document for Effluent Limitations Guidelines and Standards for the Iron and Steel Manufacturing -- Proposed, Point Source Category Vol. II, Coke Making, Sintering, Iron Making Subcategories</u>, EPA-400/1-80/024-b, December 1980b.

U.S. EPA. <u>Development Document for Effluent Limitations Guidelines and Standards for the Iron and Steel Manufacturing -- Point Source Category, Vol. III, Proposed, Steel Making Subcategory, Vacuum Degassing Subcategory, Continuous Casting Subcategory</u>, EPA-440/1-80/024-b, December, 1980c.

U.S. EPA. <u>Development Document for Effluent Limitatons Guidelines and Standards for the Iron and Steel Manufacturing -- Point Source Category Proposed, Vol. V, Scale Removal Subcategory Acid Pickling Subcategory</u>, EPA-440/1-80/024-b, December, 1980d.

U.S. EPA. <u>Environmental and Resource Conservation Considerations of Steel Industry Solid Waste</u>, EPA-600/2-79/074, PB-299919, April, 1979.

U.S. EPA. <u>Industrial Process Profiles for Environmental Use: Chapter 24. The Iron and Steel Industry</u>, EPA-600/2-77-023X, PB266226, 1977.

U.S. EPA. <u>Environmental Considerations of Selected Energy Conserving Manufacturing Process Options - Vol. III - Iron and Steel Industry Report</u>, EPA-600/7-76-034C, PB-264269, 1976.

U.S. EPA. *Development Document for Proposed Effluent Guidelines and New Source Performance Standards for Steel Making Segment of the Iron and Steel Point Source Manufacturing Category.* EPA, January, 1974.

Wilson, Jr., P.J. and Wells, J.H. (ed.) *Coal, Coke Chemicals* - *Chap. II, Chem. Eng. Series,* McGraw Hill Book Co., (1950).

CHAPTER 6

PAPER AND PULP INDUSTRY

INTRODUCTION

The pulping process is discussed in this Chapter. Pulping of wood is the initial step in the manufacture of paper and paper products. The pulping process is the conversion of fibrous raw material (wood) into a material suitable for use in paper, paperboard, and building materials. Pulp is the fibrous material ready to be made into paper.

There are four major chemical pulping techniques: (1) Kraft or sulfate, (2) sulfite, (3) semichemical, and (4) soda. Of the major pulping techniques, the Kraft or sulfate process produces over 80 percent of the chemical pulp produced annually in the United States. In 1970, there were 116 mills producing 29.6 million tons of pulp by the kraft process. During the same year, pulp and paper board consumption was 56.8 million tons [U.S. EPA, 1973].

KRAFT PULPING

The two basic components of the pulp wood are cellulose and lignin. The cellulose fibers, which constitute the pulp, are bound together by lignin. Thus, any process manufacturing pulp must remove the lignin. The Kraft process employs chemical dissolution of the lignin. The flow chart of the Kraft process is depicted in Figure 6.1.

The typical pollutants from the Kraft pulping process are hydrogen sulfate, methyl mercaptan, dimethyl sulfate, and dimethyl disulfide. The pollutants containing sulfur are collectively called total reduced sulfur (TRS). Hydrogen sulfite emissions are a direct result of sodium sulfide breakdown in the Kraft cooking liquor. Methyl mercaptan and dimethyl sulfide are formed in the reactions with lignin. Dimethyl sulfide is

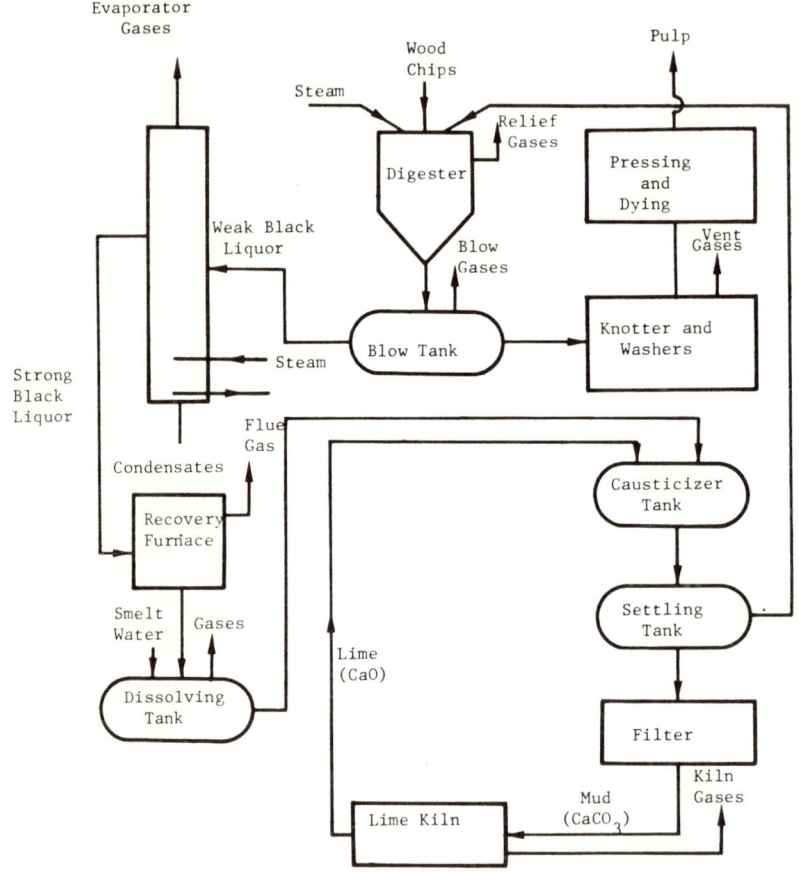

Figure 6.1. Flow Chart of Kraft Pulping Process (U.S. EPA, Sept. 1973).

formed through the oxidation of mercaptant groups derived from the thiolignin.

The Kraft process, as a whole, can be subcategorized into two parts; actual pulping, and recovery. The pulping occurs in the digester system, followed by the pulp wash. The recovery refers to multiple effect evaporators (MEE), recovery furnace, smelt dissolver, and lime kiln, all of which serve to recover the white liquor and heat.

Digester System

There are two types of digesters, continuous and batch. The composition and quality of digester gases will differ between the two digester types. Batch digesters often present air pollution problems because of gas surges produced during blowing. Continuous digesters, however, present much smaller pollution problems than batch digesters, because contaminated condensates and odorous gases flow at a regular rate. Most of the Kraft pulping is currently done in batch digesters [Goodwin, 1978]. The following discussion addresses only the batch digester system.

The digestion process is a delignification process in which white liquor and wood chips are heated to 170-175°C. White liquor is composed of sodium sulfide (Na_2S) and sodium hydroxide (NaOH).

Emissions from the digester are caused by two streams; the relief gases and blow gases. The relief gases are the ventilation gases which maintain the digester at the proper pressures (100-135 psig). The blow gases, (steam and other gases) are from the blow tank where the cooking liquor is drained from the pulp. Both the relief and blow gases are condensed for heat recovery. Table 6.1 gives the pollutant emission rates for the digester system.

Mercaptants and methyl sulfide are unavoidable by-products of Kraft pulping. However, there are a few techniques which can limit their generation. Generally, high temperatures, high sulfidity and long reaction times favor the production of the sulfur compounds [Douglas, 1966]. Thus, if the sulfidity of the white liquor is kept at a minimum, the formation of gaseous sulfur compounds will be reduced. Sarkanen, et al. (1970), suggests a 20 percent sulfidity for most paper mills.

Table 6.1 Typical Emission Rates From Batch Digester In kg Sulfur/Ton of ADP* [U.S. EPA, 1978].

Pollutant	Blow Gases	Relief Gases
H_2S	0-0.1	0-0.5
CH_3SH	0-1.0	0-0.3
CH_3SCH_3	0-0.25	0.05-0.8
CH_3SSCH_3	0-0.1	0.05-1.0

*ADP - Air Dried Pulp

Final emissions of the digester system are determined by the effective operation of the condenser (heat exchanger) units. Experience shows that the surface area of the heat exchanger, thought originally to be sufficient, may later prove to be too small [U.S. EPA, 1976]. This is due to fiber carryover by the blow gases and consequent fouling of the heat transfer surfaces. The fiber carryover is caused by excessive gas velocities at the entrance of the flow exhaust pipe. The minimum velocity to suspend solid particles is 18 fps, and the actual velocity in one blow tank reached 46 fps [Martin, 1969]. Recommendations to reduce fiber carryover include: decrease the digester wood to liquor ratio; relieve the digester to a lower pressure before blowing and/or blowing for a longer time period; installation of a cyclone separator; and providing more heat exchange surface.

Brown Stock Washer System

The washing process is a minor source of air pollution as compared to digestion, evaporation, and combustion. The emission of air contaminated with organic sulfur compounds is due primarily to the contact of air with black liquor. The amount of air and sulfur compounds ventilated depends mainly on the type of washing process and equipment. The two main washing processes are displacement and diffusion washing [U.S. EPA, 1976].

Categorically, displacement washers include vacuum washers and pressure washers. Of the two, the pressure washer's hood vent and foam tank vent will have lower flow rates and total reduced sulfur (TRS) emissions.

Typical emission rates from the vacuum washer's hood vent and the foam tank are shown in Table 6.2.

Table 6.2 Emission Rates From Vacuum Washing In kg Sulfur/Ton ADP [U.S. EPA, 1976]

Pollutant	Washer Hood Vent	Washer FoamTank
H_2S	0-0.1	0-0.01
CH_3SH	0.05-1.0	0-0.01
CH_3SCH_3	0.05-0.5	0-0.05
CH_3SSCH_3	0.05-0.4	0-0.03

Diffusion washing is superior to displacement washing, since the washing takes place in a closed reactor. In diffusion washing, there is no air involved. Thus, black liquor oxidation and odor release are very small compared to displacement washing.

Some mills employ contaminated condensates for washing. A study of 17 washing systems indicated an emission rate of 0.014 lb/ton ADP as H_2S using H_2O, and 0.35 lb/ton ADP treating with condensate [U.S. EPA, 1973]. In terms of pollution abatement, water is a better washing medium than the condensates. Another possibility is to reduce sulfur contents of the condensate by stripping.

Multiple Effect Evaporator System

Multiple effect evaporators (MEE) are utilized to concentrate weak black liquor from 12-18 percent solids to 40-55 percent solids [Goodwin, 1978]. The weak black liquor is a mixture of digester spent cooking liquor and stock washer discharge. Basically, each effect consists of a heating element and a vapor head. The hot vapors from the vapor heat of a previous effect pass to the heating element of the following effect. The vacuum is maintained by means of rapid condensation of the vapor from the final effect.

This section includes only the most important indirect steam evaporation system from the viewpoint of pollution. Thus, the following discussion is for Vacuum MEE. However, the pollution control strategies can be readily applied to other evaporation techniques. Typical emission rates for MEE are shown in Table 6.3.

Table 6.3 Emission Rates From MEE In kg Sulfur/Ton ADP [U.S. EPA, 1973]

Pollutant	Emission Rate kg Sulfur/ton ADP
H_2S	0.05-1.5
CH_3SH	0.05-0.8
CH_3SCH_3	0.05-1.0
CH_3SSCH_3	0.05-1.0

The most effective method of reduction of pollution in MEE is to steam strip the condensate, and to incinerate the noncondensable gases (H_2S, CH_3SH, CH_3SCH_3, and CH_3SSCH_3). Matteson, et al (1967) successfully stripped the condensate, generating overhead vapor of more than 95 percent hydrogen sulfide, mercaptans, and dimethyl disulfide. Steam stripping was performed in a conventional bubble cap tray stripper, resulting in a bottom product of nearly pure and reusable water. The stripped gases and other gases from the MEE can be incinerated in a lime kiln [Walther and Amberg, 1970].

The liberation of H_2S and, to a lesser extent, CH_3SH depends on the pH of the black liquor. This is due to the acidic nature of both gases. Furthermore, these gases dissociate most readily in an aqueous solution of higher pH. Addition of caustic soda to the weak black liquor can be beneficial in controlling H_2S and CH_3SH.

The emissions of H_2S and CH_3SH are also dependent on the sulfide concentration of the weak black liquor. Oxidation of this weak black liquor can convert sulfide to thiosulfate, and CH_3SH to CH_3SSCH_3. These conversions will facilitate the reduction of H_2S and CH_3SH emission from MEE. The condensate from the MEE processing of oxidized weak black liquor will require little, if any, treatment for odor abatement [EPA, 1976].

Furthermore, Galeno and Amsolen (1970) observed several benefits of weak black liquor oxidation with oxygen, as compared to lack of oxidation. The oxidation utilizing O_2 resulted in lower emissions of H_2S from the MEE, and improved the evaporator condensate water quality.

There are two major factors which should be considered in the effectiveness of weak black liquor oxidation to prevent TRS emission. First, a high degree of oxidation is required. Second, oxidized sulfur compounds have tendencies to revert to sulfide during MEE, as well as during extended storage. Long storage periods after black liquor oxidation should be avoided.

RECOVERY FURNACE SYSTEM

The recovery furnace functions include: (1) recovery of sodium and sulfur, (2) production of steam and, (3) disposition of unwanted components of the dissolved wood. The recovery furnace system generally includes the following units: recovery furnace, flue gas direct contact evaporator, primary particulate control device, and secondary particulate control device. In some cases, the direct contact evaporators have been used to contain the particulate emissions or eliminate them altogether [U.S. EPA, 1970]. Table 6.4 is a listing of emission rates from the recovery furnace.

The three direct evaporators used in the Kraft pulping mills are cascade evaporators, cyclone evaporators, and the venturi recovery unit. The difference between a venturi and the other two evaporators is its efficiency in removing particulates. While cyclone and cascade evaporators remove only 40-50 percent of the particulate matter, the venturi recovery units can be designed to capture better than 90 percent of the particulates [U.S. EPA, 1973]. The direct contact evaporator serves as an adsorption unit for SO_2 and nearly all SO_3. Furthermore, it absorbs H_2S emitted from the recovery furnace under conditions of high black liquor pH and low sodium sulfide concentrations in the strong black liquor.

Table 6.4 Emission Rates From Recovery Furnace [U.S. EPA, 1976].

Pollutants	Emission Rate kg/ton ADP
H_2S	0-25
CH_3SH	0-2
CH_3SCH_3	0-1
CH_3SSCH_3	0-0.3
SO_2	0-40
SO_3	0-4
NO_x	0.7-5

The particulate emission rates are listed in Table 6.5.

Table 6.5 Particulate Emission Rates From The Recovery Furnace [U.S. EPA, 1976].

Emission Source	Emission Rate, kg/ton ADP
Recovery System after Electrostatic Precipitator	0.5-12
After Venturi Evaporator	14-50
Flue Gas Dust Load	40-75

The direct evaporation unit is also a potential source of TRS. However, the emission of TRS depends heavily on the residual sulfide in the black liquor from MEE. For this purpose, some mills employ strong black liquor oxidation. A survey of 32 recovery furnace systems where black liquor oxidation is not used showed sulfur emission ranging from 0.75-31g/Kg ADP, with an average of 7.7g/Kg ADP. Conversely, a survey of 17 units that utilized black liquor oxidation indicated an average emission rate of 3.7g/Kg ADP [Goodwin, 1978].

It is possible to eliminate the direct contact evaporator in favor of a non-contact design. This change will cause an increase in particulate emissions, which would then require installation of a more efficient particulate control device.

The emission of sulfur compounds from the recovery furnace is also dependent on the design of the furnace. The recovery furnace has two functions; recovery of the chemicals in their reduced state (i.e., sulfur should be present as sulfide and not sulfate), and recovery of heat to generate steam for the processes.

Combustion within the recovery furnace is separated into two zones. The first zone must be maintained under reducing conditions, at less than the required stoichiometric amounts of air. This zone discharges chemicals in the molten state, with the sulfur present mainly as sulfide and organic matter as a gas having considerable heat value. The second zone of combustion starts with the addition of secondary air. The secondary air should be supplied in 10-20 percent excess of the amount required for complete combustion.

The release of sodium from the burning char of black liquor depends on the temperature and gas conditions in the border zone between the bed and the flue gas. The TRS release to the flue gas is affected by the total sulfur concentration, and the sodium to sulfur ratio. Since the sodium/sulfur ratio in the smelt bed depends on temperature, the release of sulfur to the flue gas is also a function of temperature. Hydrogen sulfide may be present in the flue gas in the primary air combustion zone at 50-100 ppm, if the zone between the bed and the flue gas is kept at the optimal temperature. Hydrogen sulfide concentrations of 15,000 ppm have been observed in this region when blackout conditions are present (blackout conditions refer to insufficient rate of combustion in the hearth). Typically, a low temperature favors the presence of sulfide and H_2S [U.S. EPA, 1976].

There are several advantages in raising the primary air temperature. The velocity, for the same flow of air, will increase and improve the sodium evaporation from the bed. Also, the release of sulfur from the bed decreases.

Immediately after introduction of secondary air, the final combustion starts. The amount of primary air for most furnaces must be more than 110 percent and less than 125 percent of the theoretical air, to avoid formation of stick dust. Stick dust has a tendency to foul the heating surfaces. The secondary air should be supplied to the furnace so that it mixes with the gas coming from the primary air combustion zone. Therefore, efficient regulation and method of introduction of secondary air is important in the generation of stick dust.

The variables affecting the TRS emissions from the recovery furnace are as follow: quantity and method of introduction of combusted air, the rate of black liquor feed, the degree of turbulance in the oxidation zone, the O_2 content of flow gas, the spray pattern and droplet size of the liquor fed to the furnace, and the degree of disturbance in the smelt bed. These variables are independent of the presence or absence of a direct contact evaporator [Theon, et al. 1968].

Teller and Amberg (1975) have developed a control technique for use in the recovery furnace which utilizes alkaline adsorption with carbon activated oxidation of the scrubbing solution. Pilot plant studies indicated that this technique is capable of reducing TRS emissions from 20-1500 ppm down to 1 to 10 ppm. This method would also alleviate particulate and SO_2 emissions [Teller and Amberg, 1975]. Another effective adsorption method is the TRS system potential, by the Weyerhaeuser Company. The TRS scrubber adsorbs up to 99 percent of H_2S and collects about 85 percent of the particulate matter.

Nitrogen oxides are generated in the recovery furnace at a rate of 0.7-5 kg/ton ADP [U.S. EPA, 1976]. The reduction of NO_x production lies in provisions for facilitation of a heat sink in the recovery furnace. The endothermic reaction of Na_2SO_4 to Na_2S in the smelt bed also acts as a heat sink to inhibit excess flame temperatures. A reducing atmosphere above the smelt bed will also reduce NO_x formation. The supply of air can be arranged to spread out the flame front and, in that way, inhibit the increase in gas temperature.

Smelt Dissolving Tank

The molten smelt accumulated in the recovery furnace as a result of combustion is dissolved in water in the smelt dissolving tank. Dissolution of the molten smelt (sodium carbonate and sodium sulfide) in water forms a green liquor. The dissolution is aided by agitators and steam, or a liquid shatterjet system, to break up the smelt stream before it enters the solution. The large volume of steam generated by the contact of smelt with water is vented. Table 6.6 depicts the emission rates from the smelt dissolving tank.

Table 6.6 Emission Rates From Smelt Dissolve Tank [U.S. EPA, 1976]

Pollutants	Kg Sulfur/ton ADP
H_2S	0-1.0
CH_3SH_3	0-0.8
CH_3SCH_3	0-0.5
CH_3SSCH_3	0-0.3
	Kg/ton ADP
Particulate Emission (After Control Device)	0.01-0.5
SO_2	0-0.2

Particulate matter contains NaOH, Na_2CO_3, and Na_2S. The particulate emissions are captured from escaping vent gases by mist eliminator pads. Typical efficiency of the pads for particulate removal is 70-90 percent. However, higher efficiencies can be achieved by facilitating the smelt tank with a spray or packed scrubber in addition to mist eliminator pads. Combinations of this type can be 98 percent efficient in removing particulate matter [U.S. EPA, 1976]. Another alternative is to combine vent gases from the smelt tank with flue gases from the recovery furnace prior to entering the particulate collection device. However, the effectiveness of a electrostatic precipitator is reduced due to the water vapor content of the smelt tank vent gases. Furthermore, the Na S entrained in the smelt tank vent gases coming into contact with CO_2 from the recovery furnace may promote formation of H_2S. It appears that the best particulate control device for the smelt tank is a combination of mist eliminator pads and a scrubber.

The TRS emissions depend on the sulfide content of particulate matter, the turbulence in the dissolving tank, the type of solution used in a scrubber, if present, and pH of the scrubber liquor. Fresh water is the best solution for scrubbing and is capable of producing 0.001 lb sulfur/ton ADP [Martin, 1969].

LIME KILN

The green liquor is converted to white liquor in the lime kiln. This unit is an essential element of the closed-loop system. The kiln calcines the calcium carbonate which precipitate from the causticizer, to produce quicklime (CaO). The calcined calcium carbonate precipitate is also called lime mud. The quicklime is wetted (slacked) by the water in the green liquor solution, to form calcium hydroxide for the causticizing reaction.

The mud is contacted by the hot gases produced by the combustion of natural gas or fuel oil, and proceeds through the kiln in the opposite direction of the gas flow. The lime mud is a 55-60 percent solid-water slurry and is fed at elevation to the kiln. In the upper part of the kiln the mud dries, while at the lower end (the high temperature zone 1800-2000°F), it agglomorates into small pellets and is calcined to CaO. Typical emission rates from lime kiln are given in Table 6.7.

Variables affecting the TRS emission of the lime kiln include the temperature at the cold end of the kiln, the O_2 content of the gases leaving the kiln, the sulfide content of the lime mud, and the pH and sulfide content of the water used in the particulate scrubber [NCASI, 1971]. If contaminated condensate is used as the scrubbing solution, the exhaust gases could strip out the dissolved TRS and increase the TRS emission from the lime kiln.

Table 6.7 Lime Kiln Emission Rates [U.S. EPA, 1976]

Pollutants	Lime Kiln Exhaust & kg Sulfur/ton ADP	Lime Slaker Vent kg Sulfur/ton ADP
H_2S	0-0.5	0-0.1
CH_3SH	0-0.2	0-0.01
CH_3SCH	0-0.1	0-0.01
CH_3SSCH_3	0-0.05	0-0.01
	kg/ton ADP	kg/ton ADP
SO_2	0-1.4	--
NO_x	10-25	--

The two most important preventive measures in terms of TRS emission are; maintenance of the proper process conditions, and scrubbing the exhaust gases with caustic solution. For example, TRS emission can be reduced by a sufficient supply of O_2 and a reduction in sulfide content of the mud.

The lime kiln is operated with excess air and high temperature. Both conditions favor the production of NO_x. Therefore, any measure to reduce TRS may promote generation of NO_x. Thus, oxides of nitrogen are unavoidable by-products of the lime kiln and any measure to control their release should occur after the kiln.

Table 6.8 summarizes some of the pollution problems associated with the Kraft pulping process, and suggests control alternatives.

RECOMMENDED AREAS FOR PROCESS MODIFICATION RESEARCH

After evaluating the process technology, the following areas are recommended for further

Table 6.8. Pollutants and Control Options in Kraft Pulping Process

Process	Pollutants	Sources in Process	Nature of Pollutants	Pollutant Control Strategy
Digester	TRS, relief & blow gases	Digester Tank blow tank	Organic, gases	Lower sulfidity, lower blow pressure, reduce fiber carryover by cyclone or other methods.
Brown Stock Washer	TRS, noncondensables	Vacuum washer	Organic, gases	Use water as washing fluid, employ diffusion washers.
Multiple Effect Evaporators	TRS, noncondensables	Evaporators	Organic, gases	Treatment of weak black liquor by oxidation and with caustic soda. Effective stripping of condensate from the condenser unit.
Recovery Furnace System	TRS, SO_2, SO_3, NO_x Particulates	Recovery furnace direct contact evaporator	Organic and inorganic gases and solids	Strong black liquor oxidation. Use direct contact evaporator, venturi type. Improve design. Alkaline adsorption with carbon-activated oxidation of the scrubbing solution.
Smelt Dissolve Tank	TRS, SO_2, Particulates	Smelt Tank	Organic and inorganic gases and solids	Combination of mist eliminator pad and scrubber unit. Use fresh water for scrubbing.
Lime Kiln	TRS, SO_2, NO_x	Kiln	Organic and inorganic gases	Maintain proper process conditions. Sufficient supply of O_2 and reduction in sulfur content of the mud.

research and development:

- Research and digestion of wood chips to determine optimal sulfidity, pH, and temperature to reduce pollution.

- Studies on control of fiber carryover from the blow tank by cyclone and/or by reduction of relief pressure and selection of wood to liquor ratio to prevent TRS emission.

- Comparative studies on design and application of diffusion and displacement washers to minimize TRS.

- Effect of black liquor oxidation and pH control on weak and strong black liquor to reduce TRS emission.

- Studies on the direct and non-contact evaporation in the recovery furnace system, to determine capability and merits of each unit in reducing the TRS emission.

- Research on combustion of strong black liquor to prevent blackout conditions and stick dust formation, by optimizing the design and operating conditions in the recovery furnace.

- Research and application of scrubbing techniques to reduce TRS emission.

REFERENCES

Douglas, I.B., "The Chemistry of Pollutant Formation in Kraft Pulping." In proceedings of International Conference on Atmospheric Emissions from Sulfate Pulping, April 28, 1966. E.O. Painter Printing Co., 1966.

"Factors Affecting Emission of Odorous Reduced Sulfur Compounds from Miscellaneous Kraft Process Sulfur", NCASI Technical Bulletin, No. 60, March 1972.

Galeano, S.F. and Amsden, C.D. "Oxidation of Weak Black Liquor with Molecular Oxygen." TAPPI, Vol 53, November 1970, pp. 2142-2146.

Goodwin, D.R. "Draft Guideline Document: Control of TRS Emissions from Existing Kraft Pulp Mills", EPA 450/2-78-003A. January 1978.

Hansen, S.P. and Burgess, F.I. "Carbon Treatment of Kraft Condensates Wastes." TAPPI, Vol. 51, June 1968, pp. 241-245.

Martin, G.C. "Filter Carryover With Blow Tank Exhaust", TAPPI, Vol. 52, No. 12, December 1969.

Matteson, M.J., et al. "Sekor II: Steam Stripping of Volatile Substances from Kraft Pulping Mill Effluent Stream." TAPPI, Vol 50, No. 2, February 1967.

Prakash, C.B. and Murry, F.E. "Studies on H_2S Emissions during Calcining", Pulp and Paper Magazine of Canada, Vol. 74, May 1973, pp. 99-102.

Sarkanen, K.V., et al., "Kraft Odor", TAPPI. Vol. 53, No. 5, May 1970.

Suggested Procedure for the Conduct of Lime Kiln Studies to Define Emissions of Reduced Sulfur Through Control of Kiln and Scrubber Operating Variable." NCASI Special Report No. 70-71, January 1971.

Teller, A.J. and Amberg, H.R. "Considerations in the Design for TRS and Particulate Recovery from the Effluents of Kraft Recovery Furnace." Preprint TAPPI Environmental Conference, May 1975.

Theon, G.N., et al. "The Effect of Combustion Variable on the Release of Odorous Sulfur Compounds from a Kraft Recovery Furnace, TAPPI, Vol. 51, No. 8, August 1968, pp. 324-333.

U.S. EPA. "Atmospheric Emissions From the Pulp and Paper Manufacture Industry." Environmental Protection Agency. Research Park N.C. Office of Air Quality Planning and Standards. EPA 459/1-73/002, September 1973.

U.S. EPA. "Control of Atmospheric Emissions in the Wood Pulping Industry", Environmental Engineering Inc., and J.E. Sirrine Company, Final Report, EPA Contract No. CPA-22-69-18, March 15, 1970.

U.S. EPA. "Environmental Pollution Control Pulp and Paper Industry: Part I Air", EPA 625/7/76-001, October 1976.

Walther, J.E. and Amberg, H.R. "A Positive Air Quality Program at a New Kraft Mill,, *Journal of Air Pollution Control Assoc.*, Vol. 20, No. 2, January 1970.

Walther, J.E. and Amberg, H.R. "The Role of the Direct Contact Evaporator in Controlling Kraft Recovery. Furnace Emissions", *Pulp and Paper Magazine of Canada*, Vol. 72, October 1971, pp. 65-67.

SUPPLEMENTAL REFERENCES

Anderson, H.K. and Ryan, J. "Improved Air Pollution Control for a Kraft Recovery Boiler: Modified Recovery Border No. 3", EPA 650/2-74-071-a, August, 1974.

Bhatia, S.P., et al., "Removal of Sulfur Compound from Kraft Recovery Stack with Alkaline Suspension of Activated Carbon", *TAPPI*, Vol. 56, December 1973, pp. 164-167.

Clement, J.L. and Eliot, J.S. "Kraft Recovery Boiler Design for Odor Control", *Pulp and Paper Magazine of Canada*, Vol. 78 No. 8, February 7, 1969, pp. 47-52.

Cooper, H.B.H. and Rossano, A.T., Jr. "Black Liquor Oxidation with Molecular Oxygen in Plug Flow Reactor", *TAPPI*, Vol. 56, June 1973, pp. 100-103.

Douglas, I.B. "Sources of Odor in the Kraft Process: Odor Formation in Black Liquor Multiple Effect Evaporators." *TAPPI*, Vol. 52, September 1969, pp. 1738-1741.

CHAPTER 7

THE PRIMARY ALUMINIUM INDUSTRY

INTRODUCTION

The primary aluminium industry consists of processing bauxite ore to produce alumina (and occasionally aluminium hydroxide), and processing the alumina to produce aluminium. Approximately, 7.6×10^6 tons of alumina were produced in U.S. from processing about 15.4×10^6 tons of bauxite in 1972 and 94 percent of the alumina was utilized to make aluminium [Saxton and Kramer, 1975].

BAUXITE PROCESSING

Bauxite is composed mainly of metallic oxides, with aluminium oxide comprising from 20 to 60 percent of the ore as mined. Approximately 87 percent of the bauxite ore is imported, mainly from Jamaica and Surinam. The environmental problems associated with bauxite mining and shipping are caused by fugitive dust and water runoff. Bauxite ore may contain up to 30 percent moisture, and should be dried at 110 °C before shipping. An overview of the bauxite processing is shown in Figure 7.1.

The raw bauxite ore is benefacted by grinding to about 100 mesh, digested in caustic at elevated temperatures and pressures (145 °C and 60 psig), and then filtered or thickened in the presence of flocculants. These operations yield a sodium aluminate liquor, and a waste stream of "red mud" which contains the impurities found in the bauxite ore.

The aluminate liquor is diluted and cooled to hydrolyze the sodium aluminate, forming a precipitate of aluminium hydroxide which is filtered and calcined to alumina. The alumina is then shipped to aluminium smelting. Depending on the type of bauxite processed, 1/3 to 2 kg of red mud is produced per each kilogram of alumina product. Table

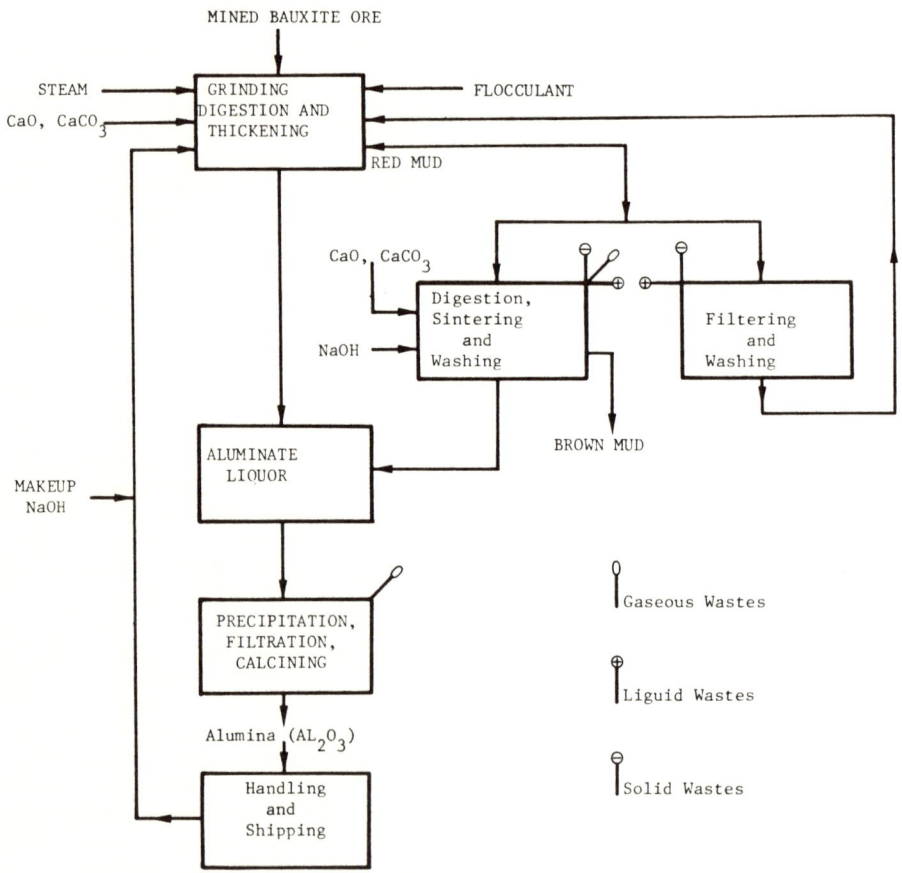

Figure 7.1. Bauxite Processing

7.1 gives the chemical analysis of red mud slurries from different bauxite ores [U.S. EPA, 1974]. The pH of this slurry is approximately 12.5.

Table 7.1 Chemical Analysis of Red Muds

Component	Weight %		
	Arkansas	Surinam	Jamaica
Fe_2O_3	55-60	30-40	50.54
Al_2O_3	12-15	16-20	11-13
SiO_2	4- 5	11-14	2.5- 6
TiO_2	4- 5	10-11	—
CaO	5-10	5- 6	6.5-8.5
Na_2O	2	6- 8	1.5-5.0

In the combination process, used for bauxite ores with high silica content such as those from Arkansas, the red mud residue is treated to extract additional amounts of alumina and to recover sodium values. This additional extraction step is accomplished by mixing red mud with limestone and soda ash, and then sintering this mixture at 1100 to 1200° C. The important reactions are the conversion of silica to calcium silicate, and residual alumina to sodium aluminate. The sintered products are leached to produce additional sodium aluminate solution, which is either filtered and added to the main stream for precipitation or precipitated separately. By the addition of seed material and careful control of composition and agitation, alumina trihydrate is precipitated.

The ideal solution to the red mud problem would be to develop a use for it. A possible application utilizes the high iron content of the red mud (Table 7.1). Fursman, et al. (1970) described a process based on sintering the red mud with carbon and limestone, and melting the sinter in an electric ore furnace to produce a low grade iron which could be further processed into steel.

This process was further developed [Guccione, 1971] but has not yet found commercial application. Other investigators have examined the applicability of red mud to manufacture portland cement, brick, and road construction [Fursman, et al. 1970; Solyman and Briudoso, 1973].

There is also a need to develop an economically viable leaching and extraction process for the recovery of mineral values from red mud wastes.

During the handling and shipping of alumina, dust is formed. The tendency of aluminas to form dust during handling depends on the degree of calcination and on the particle size distribution. The size distribution and adsorption capability of calcined alumina are important factors in the reduction and dry gas cleaning stages of aluminium manufacture, as will be discussed later.

With highly calcined, flowing aluminas, the formation of dust is prevented by the surface roughness, which increases as the α-alumina content rises. If fine grained hydroxide is calcined to a lower degree, it will start developing dust. The tendency of weakly calcined aluminas to cause dusting is counteracted by minimizing the amount of fines. This is normally done in the precipitation area of the process, where the particle size distribution of the hydroxide is controlled.

Decisive for the mechanical strength acquired by the individual hydroxide particles are [Schmidt, 1980]:

- The basic principle applied for crystallization, i.e., nucleation, crystal growth, or agglomeration conditions.

- Influence of liquor impurities such as $CaCO_3$, oxalates etc.

- The mechanical stresses to which the hydroxides are exposed during precipitation.

Schmidt, et al. (1980) reported that the gas/solid velocities and thermal stresses in the fluid bed calciner and in the rotary kiln have a strong influence on the particle size distribution of the calcined alumina.

Thus, a research program to improve the mechanical strength, adsorption capability and particle size distribution of calcined alumina by modifying the precipitation and calcining of alumina is indicated.

Studies conducted in Russia and elsewhere indicate the possibility of using nitric acid stripping for the extraction of alumina from ores containing high amounts of silica. Sutyrin and Zverev (1976) reported that stripping at 160°C lowered acid usage and produced solutions less contaminated by iron. Acid stripping of domestic U.S. ores which contain high amounts of silica should be optimized by proper control of the stripping temperature.

The enrichment of high silica bauxite ores using heterotrophic bacteria is reported by Andreev, et al. (1976). It was possible to produce a concentration with 48.4 percent Al_2O_3 content at a recovery of 74 percent. Thus, additional research is indicated, to develop acid stripping and bacterial enrichment techniques for domestic bauxite ores.

PRIMARY ALUMINIUM SMELTING

The large scale, economical production of primary aluminium became possible when, in 1886, C.M. Hall and P. Heroult independently developed the electrolytic process which has remained essentially unchanged, except for improvements in equipment design and operating practices. It is used in all the commercial processes in the U.S. producing primary aluminium. There are 31 aluminium reduction plants in the U.S., with a total annual capacity of about 4.5×10^6 tons. The energy consumed annually at full production is estimated to be in the range of 80 to 100 billion K.W.H.

The basic process for reducing alumina to aluminium is shown schematically in Figure 7.2. The raw materials used in aluminium production include alumina, cryolite (a double flouride of Na and Al), pitch, petroleum coke, and aluminium flouride. For every kg of aluminium produced, about 2 kg alumina, 0.25 kg pitch, 0.05 kg cryolite, 0.5 kg petroleum coke, 0.04 kg aluminium flouride, 0.6 kg

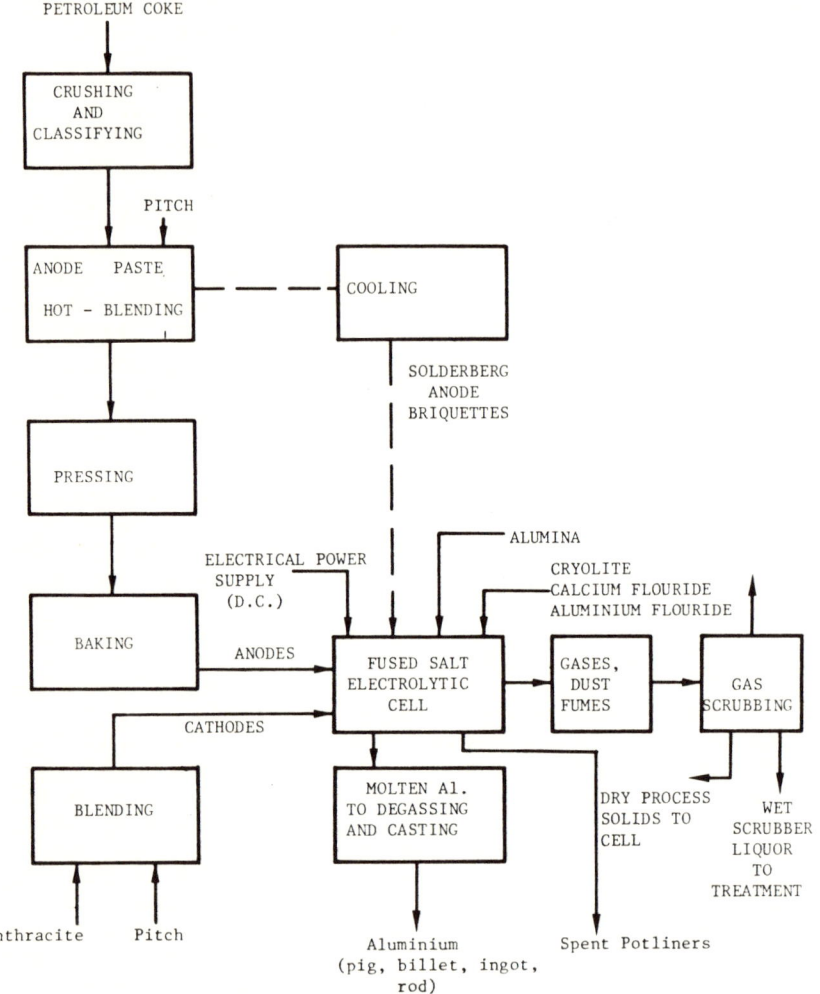

Figure 7.2. Primary Aluminium Production.

baked carbon and 22 K.W.H. of electrical energy are needed.

The heart of the aluminium plant is the electrolytic cell, which consists of a steel container lined with refractor brick, with a bottom inner lining of carbon. The cells are arranged in rows in an operating unit (potline) and as many as 100 to 250 cells are electrically connected in series. The electrical supply is direct current, and is on the order of several hundred volts and 60,000 - 100,000 amperes. The carbon lining at the bottom of the cell acts as a cathode when covered with molten aluminium. The anode of the cell is baked carbon. The electrolite consists of a mixture of cryolite (80-85 percent by wt.), calcium flouride (5-7 percent), and alumina (2-8 percent). Alumina is added to the bath intermittently, to maintain the concentration of dissolved alumina within the desired range. The fused salt bath is usually at a temperature of 900° C.

The reaction in the aluminium reduction cell is not completely understood [Kirk-Othmer, 1963]. Apparently aluminium is reduced from the trivalent state (assuming ionization in the molten salt), to the liquid metal state at the cathode. Oxygen, assumed present in the bath in the divalent state, appears at the carbon anode and immediately reacts with the anode and forms a mixture of CO_2 (≈ 75 percent) and CO (≈ 25 percent).

Thus, the operation of the electrolytic aluminium reduction cell results in the continuous consumption of alumina and the carbon anode, and the evolution of gaseous reaction products. The aluminium is withdrawn intermittently from the bottom of the molten bath, and is collected in ladles and cast into ingots.

The various anode making operations are conducted in the anode paste plant. The anode paste consists mainly of high grade coke (petroleum and pitch coke) and pitch, with a maximum of 0.7 percent ash, 0.7 percent sulfur, 8 percent volatiles, 0.5 percent alkali and 2 percent moisture. The two types of pitch handling systems are, solid pitch handling and liquid pitch handling (using organic liquids as a heat transfer medium). Dust and toxic emissions are the main pollutants from these systems.

The two different anode systems used differ in the replacement of the anode.

1. Prebaked system - the anode is replaced intermittently,

2. Solderberg system - the anode is replaced continuously by the anode paste descending from the anode shell suspended above the electrolytic cell.

The carbon cathode has an average life of between 2-3 years. The spent cathode lining is removed by drilling and/or soaking in water. About 1200 m^3 of shell waste is generated each year in the U.S.

The major pollutants generated during the smelting of alumina include particulate emissions, inorganic fluorides, oxides of sulfur, H_2S, carbon disulfide, carbonyl sulphide, CO and CO_2, and spent carbon cathode.

Emissions from the horizontal-stud Solderberg cell and vertical-stud Solderberg cell [Kotari, et al. 1974] are shown in Table 7.2

Table 7.2   Emissions From Solderberg Cell

Cell Type	Emissions	lb. per ton of Aluminium
Vertical-stud	Particulates	98.4
	Gaseous Fluorides (HF)	26.6
	Particulate Fluoride	15.6
Horizontal-stud Solderberg	Particulates	78.4
	Gaseous Fluorides (HF)	30.4
	Particulate Fluoride	10.6

The particulates contain Al_2O_3, Na_2CO_3 and carbon dust. Approximately 60 percent of the particulates are less than 5 μm in size. Increasing the mechanical strength of Al_2O_3 by modifying precipitation and calcination operations (as discussed earlier) may lead to a reduction of particulate Al_2O_3 emissions.

The emissions of both particulate and fluorine compounds from the bath increase with increasing temperature, decreasing alumina content, and decreasing bath ratio of NaF/AlF$_3$ [Kotari, et al. 1974]. The hydrogen fluoride emissions are generated primarily as a result of moisture treating with AlF containing materials. The HF emissions increase directly with the water content [Bell and Dawson, 1971]. The sources of moisture are the atmosphere, and the moisture content of Al_2O_3.

The fluoride particulates range in size from about 0.05 - 0.75 μm, with the majority of the particles smaller than 0.25 μm [McCabe, 1975]. The fluorine content of the total gases, withdrawn from the pots or pot rooms may vary from 2 to 40 mg/ft^3, or between 18.97 - 25.63 kg/ton of the raw material [Kotari, et al. 1974]. Cooling the process by 5°C will reduce fluorine consumption by 0.2 kg/ton of raw material [Kotari, et al. 1974]. An optimizaton of the operative temperature of the electrolytic cell is needed to minimize the generation of hydrogen fluoride. Fifteen percent of the fluoride present in the emissions from prebaked pots occurs as HF, while 90 percent of the flouride in Solderberg pot emissions occurs as HF [Less and Waddington, 1971].

Henry (1963) reported on experimental work which established correlations between three cell operating parameters and effluent quality. The results of these experiments are summarized in Table 7.3. It was thus shown that by increasing he bath ratio (NaF/AlF$_3$) and the alumina content of the bath, and decreasing the cell temperature, a decrease was seen in the fluoride content of the cell effluent. Efforts should be directed to evaluate these and other cell operational variables, in order to minimze fluoride release.

There have been several research efforts directed at adsorbing HF on alumina. Investigations performed under production conditions [Colpitts, 1972; Chavinean and Muhlrad, 1973; Cochran, 1974] show that the efficiency of the adsorption process falls off once a certain quantity of gaseous fluorine has been adsorbed, and this quantity is directly proportional to the specific surface area of Al_2O_3. In a recent study, Baverez and DeMarco (1980) showed that HF is adsorbed on Al_2O_3 as two bimolecular layers.

Table 7.3 Effect of Cell Operating Parameters as Fluoride Effluent

Range of Variables			Effect on Fluoride Effluent
Cryolite Bath Ratio (NaF/AlF)	Alumina Content	Temp. °C	
1.44 to 1.54	4%	975°C	31% decrease
1.50	3% to 5%	975°C	20% decrease
1.50	4%	982°C	24% decrease

In other studies on HF control by adsorption onto aluminum oxide, it was found that the maximum adsorption capacity determined in the laboratory is roughly twice the observed under produciton conditions. Research efforts are needed to explain this discrepancy between laboratory and production tests and thereby improve the adsorption of HF on Al_2O_3 in industrial pollution control devices. Any increase in the surface area of Al_2O_3 per unit weight would increase its HF adsorption capacity.

Sulfur oxides in the fumes from electrolytic cells are generally removed by wet scrubbing (in lime scrubbers) and/or by dry scrubbing (by adsorption on Al_2O_3). It is reported [U.S. EPA, 1973] that a venturi type lime scrubbing system at the Mitsui Aluminium Company, Ltd. has operated at SO_2 removal efficiencies of 86-93 percent for more than a year. It may be necessary to improve on the venturi contacting device to improve SO_2 removal efficiencies, or to use additional dry scrubbing units to further remove SO_2.

The low equilibrium value of SO_2 adsorption on smelter grade Al_2O_3 is a limiting factor in the removal of SO_2 from cell gases in a dry scrubber. It was shown by Lamb (1979) that the presence of adsorbed fluoride from cell gas would reduce equilibrium adsorption of SO_2 even further. This effect will be most important with vertical stud Solderberg cells, where HF loading and concentration is higher than the horizontal stud Solderberg or prebaked electrolytic ells. It is necessary to understand the interaction of SO_2 and HF on Al_2O_3

in order to develop dry scrubber systems to more efficiently remove SO_2.

Tschopp, Franke and Bernhauser (1979) report that additions of lithium reduced the operating temperature of the elctrolytic cell by 14°C and reduced fluorine evolution by 25 percent, based on their studies at a Swiss aluminium plant at Essen.

The used cell linings consist largely of carbon (the cathode), but also contain cryolite, aluminium carbide, and aluminium nitride. NaOH and sodium cyanide also appear in waters resulting from leaching of cell linings. The carbides react with water vapor in the air to produce methane (CH_4), hydrogen, C_2H_2 and other hydrocarbons. The nitride yields ammonia. This cathode material, in the past, had been dumped or used as landfill. However, environmental regulators now consider the fluoride content to pose a possible pollution hazard. Over 30 percent of the fluoride in scrapped cell linings is water soluble. Lu and Shelley (1970) describe eleven treatment methods to treat cell linings. They concluded that apart from the two methods (sodium hydroxide leaching and steam hydrolysis) already in commercial use, the most promising approach seems to be the use of hot water hydrolysis and recycle of treated cathode material. However, since it is unlikely that all the cathode carbon can be recycled, the problem of fluorides and cyanides in scrap cathode still remains. It is necessary to develop leaching methods to remove fluorides and cyanides, to facilitate disposal of cathode linings.

Table 7.4 lists the major pollution problems associated with the Primary Aluminium Industry.

RECOMMENDED RESEARCH STUDIES FOR POLLUTION CONTROL IN THE PRIMARY ALUMINIUM INDUSTRY

After evaluating the process technology, the following areas are recommended for further research and development.

- Leaching and extraction of red mud to recover mineral values.

Table 7-4. Suggested Process Modifications for Pollution Control in the Primary Aluminium Industry.

Process	Pollutants	Source in Process	Nature of Pollutants	Suggested Process Modifications
Bauxite Processing	Red Mud	Ore Digestion	Fe_2O_3, Al_2O_3, etc. oxides	Recovery of mineral values, Developing alternative Uses for red mud.
Handling and Shipping	Alumina	--------	Particulates	Optimizing precipitation and calcination to improve mechanical strength and particle size distribution of alumina.
Primary Aluminium Smelting	Alumina	Electrolytic Cell	Particulates	Optimizing precipitation and calcination to improve mechanical strength and particle size distribution of alumina.
Primary Aluminium Smelting	Flourides	Electrolytic Cell	HF	Improving calcination of Al_2O_3 to decrease moisture content; optimizing cryolite bath ratio, alumina content and temperature of cell; use of lithium as additive; and, increasing adsorption capacity of Al_2O_3.

Table 7-4. Suggested Process Modification for Pollution Control in the Primary Aluminium Industry. (Cont'd)

Process	Pollutants	Source in Process	Nature of Pollutants	Suggested Process Modifications
Primary Aluminium Smelting	SO_x	Electrolytic Cell	----------	Improving adsorption of SO_x on Al_2O_3 in the presence HF; Use of additives to remove sulfur in coal as slag.
Primary Aluminium Smelting	Spent Cathode Linings	Electrolytic Cell	Carbon, cryolite, aluminium carbide, aluminium nitride, cyanides and flourides	Leaching to remove cyanides and flourides; and hot-water/steam hydrolysis and sodium hydroxide leaching to recycle spent cathode.

- Improving the mechanical strength, adsorption capacity and particle size distribution of calcined alumina, by optimizing precipitation and calcination of alumina.

- Optimizing the temperature for and stripping of high silica U.S. bauxite ores.

- Enrichment of high silica U.S. bauxite ore using bacterial action.

- Improving calcination of Al_2O_3 to reduce its moisture content, thereby reducing formation of HF in electrolytic cells.

- Optimization of the cryolite bath (NaF/AlF_3) ratio, alumina content and cell temperature, to reduce fluoride emissions.

- Increasing the adsorption capacity of Al_2O_3 to remove HF in the fumes from electrolysis.

- Understanding the interaction of HF and SO_x on Al_2O_3, to improve dry scrubbing efficiency of SO gases from the electrolytic cell.

- Effect of adding lithium on cell operating temperature and HF emissions.

- Leaching of cathode linings to remove fluorides and cyanides.

REFERENCES

Andreev, P.I., et al. Light Metal Age, 34 (3-4), 5-6, April, (1976).

Ball, D.F. and Dawson, P.R. Air Pollution from Aluminium Smelters, Chem. Proc. Eng., 52, (1971).

Baverez, M. and Demarco, R. J. of Metals, 32(1), p. 10-14, Jan. (1980).

Chavinean, A. and Muhlrad, W. Dry Process for Control of Fluorine Compounds Released by Alumina Reduction Pots, Paper Presented at AIME Annual Meeting, Chicago, Illinois, (1973).

Cochran, C.N. Environ. Sci. Tech., 8 (1), p. 63-66. (1974).

Colpitts, J.W. Evolution of Aluminas for the Alcoa 398 Process, Paper Presented at AIME Annual Meeting, San Francisco, CA, (1972).

Fursman, et al. Utilization of Red Mud Residues from Alumina Production, U.S. Dept. of the Interior, Bureau of Mines, Report of Investigations, No. 7454, (1970).

Guccione, E. Red Mud, a Solid Waste Can Now Be Converted to a High-Quality Steel, Eng. & Mining J., 172, 9, 136-7, (1971).

Henry, J.L. A Study of Factors Affecting Fluoride Emissions from IOKA Experimental Aluminium Reduction Cell, in "Extractive Metallurgy of Aluminium", Vol. 2, pp. 67-81, Interscience Publ., New York, (1963).

Kirk-Othmer, Encyclopedia of Chemical Technology (2), Interscience, (1963).

Kotari, V., et al. Trace Pollutant Emissions from the Processing of Metallic Ores, U.S. EPA/650/2-74/115, PB-238 265, Oct., (1974).

Lamb, W.D. J. of Metals, 31(9), pp. 32-37, Sept., (1979).

Less, L.N. and Waddington, J. The Characterization of Aluminium Reduction Cell Fume, AIME, New York, (1971).

McCabe, L.C. Atmospheric Pollution, I&EC, 47(8), August, (1955).

Saxton, J. and Kramer, M. EPA Finding on Solid Wastes from Industrial Chemicals, April 28, (1975).

Schmidt, H.W., et al. J. of Metals, Vol 32(2), 31-39, Feb., (1980).

Solyman, K., and Brijdoso, E. Properties of Red Mud in the Bayer Process and Its Utilization, Paper No. A73-56, Metallurgical Society of AIME, Feb. 28 - Mar. 1, (1973), Chicago, Illinois.

Sutyrin, Yu. E. and Zerev, L.V. Light Metal Age, 34 (1-2), 9-10, Feb., (1976).

Tschopp, Th., Franke, A. and Bernhauser, E. J. of Metals, 31 (6), pp. 133-135, June, (1979).

U.S. EPA. Operation and Performance of the Lime Scrubbing System at Mitsui Aluminium Co., Ltd. U.S. EPA 450/2-73-007, Dec., (1973).

U.S. EPA. Trace Pollutant Emission from the Processing of Metallic Ores. EPA 650/2-74-115, Oct., (1974).

West, C.E. Light Metal Age, 38 (9-10), pp. 16-18, October (1980).

CHAPTER 8

PHOSPHATE FERTILIZER INDUSTRY

INTRODUCTION

Fertilizers in general can be categorized by their composition of plant nutrients. The fertilizers differ primarily in their composition of nitrogen, phosphorous, and potassium. Normal superphosphate contains only one nutrient, phosphorous. Ammonium phosphate contains phosphorous and nitrogen. Generally, the solid and liquid mix fertilizers contain all three nutrients, in varying amounts.

Over 44 million metric tons of phosphate rock were mined in the United States during 1975. Approximately 22.75 million metric tons were consumed by the fertilizer industry during the same period [Nyers, et al. 1979]. The phosphate based fertilizers are produced by conversion of insoluble phosphate ore into the soluble form necessary for plant consumption. The phosphoric acid, backbone of phosphate fertilizer, is formed by mixing phosphate rock with sulfuric acid.

This chapter concentrates on the production of phosphoric acid, normal superphosphate, and ammonium phosphate. The preparation of phosphate rock is not discussed.

WET PROCESS PHOSPHORIC ACID PRODUCTION

Most of the phosphoric acid in the fertilizer industry is prepared by the wet process. In this process, phosphate rock reacts with sulfuric acid to form phosphoric acid and gypsum. The overall chemistry of this reaction is shown in equation 8.1 [Corbridge, 1978]:

$$Ca_{10}(PO_4)_6F_2 + 10\ H_2SO_4 + 20H_2 \rightleftarrows$$
$$6H_3PO_4 + 2HF + 10CaSO_4 + 2H_2$$
$$\text{(gypsum)}$$

(8.1)

Phosphoric acid is the most important intermediate in the production of phosphate fertilizer. In addition to its use in the production of ammonium phosphate and concentrated superphosphate, it is an intermediate for mixed fertilizers, both liquid and solid [Nyers, et al. 1979].

A simplified flow diagram of the wet process for phosphoric acid production is given in Figure 8.1. The process consists of three steps: (1) reaction of phosphate rock, (2) separation of acid from the sulfate, and (3) concentration of the acid. Important variables for a successful operation are type of phosphate rock used, the temperature of the reaction, and the slurry density as controlled by recycling weak acid to the digester step.

Most wet process plants employ the crystallization of calcium sulfate in the dehydrated state, as opposed to hemihydrate or anhydrate. This is due to the fact that the dehydrate process does not impose as many operating problems as the hemihydrate or anhydrate process. Acid produced by the dehydrate process contains about 13 percent phosphorus (30 percent as P_2O_5) and is generally concentrated to 17 to 24 percent P (40 to 54 percent P_2O_5) before use. The concentration of phosphoric acid is facilitated by continuous heated vacuum evaporators [Olson, et al. 1971].

There are substantial amounts of pollutants evolved by the wet process. A large portion of reactants become by-products contributing to the pollution. The main by-product of the dehydrate process is the impure gypsum, of little practical use. Phosphate rock is composed of 2/3 gypsum, thus making the disposal of the by-products a formidable problem. The other by-product is fluoride, which evolves during digestion of phosphate rock and can be recovered with water by a relatively simple scrubbing procedure.

The evolution of gaseous fluoride is due to the vapor pressure of hydrogen fluoride. The emission rate varies with temperature, concentration, absolute pressure, and exposed area of the liquid surface on the digester. Fluoride emissions include silicon tetrafluoride, which is formed by the reaction of hydrogen fluoride in the reactor. Tetrafluoride formation is favored at temperature

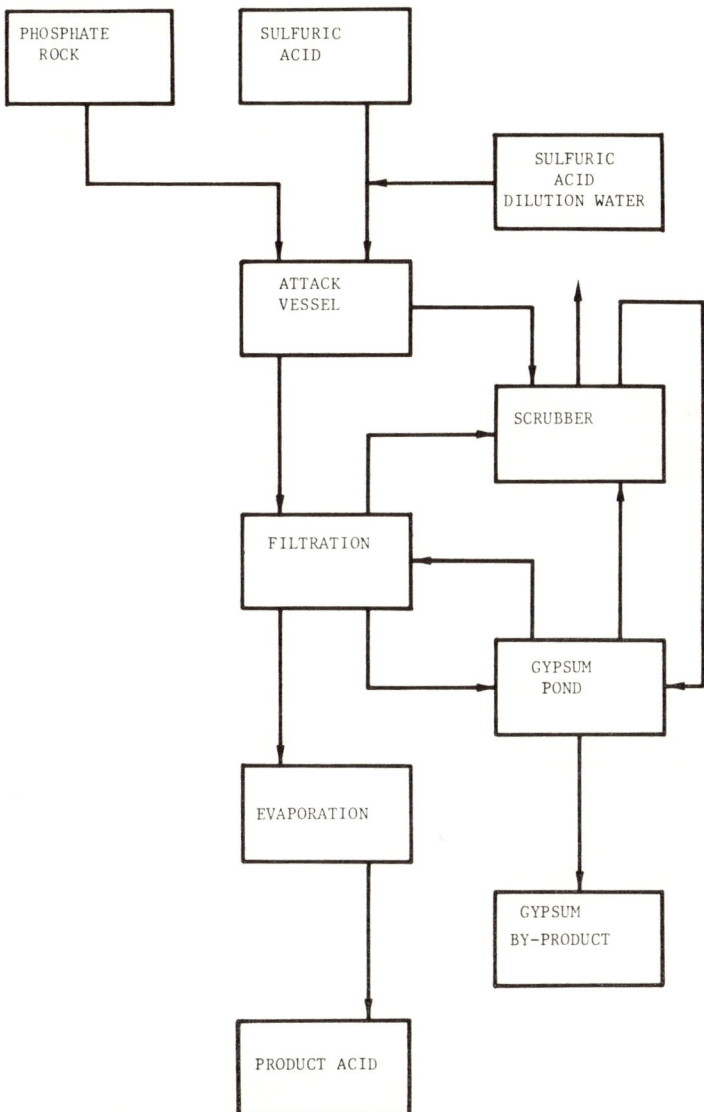

Figure 8.1. Wet Process Phoric Acid Production Process (Nyers, et al. 1979).

below 100°C [Nyers, et al. 1979].

Poorly controlled wet process phosphoric acid (WPPA) plants may have emissions as high as .07 pound of fluoride/ton of P_2O_5. On the other hand, well controlled plants using packed scrubbers or other equally effective control devices can achieve fluoride emissions below .02 pound/ton of P_2O_5 input [U.S. EPA, 1974]. Data indicate that the emission factors for plants that use fluoride recovery are not significantly different from those plants that do not employ recovery. The recovery refers to fluorine, recovered as fluorosilicic acid, fluorides, fluorosilicates, or other by-products [Nyers, et al. 1979].

The final fluorine emission is strongly dependent on the type of scrubber used, scrubber operation, and use of fresh water tail gas scrubbers. Plants that use fluorine recovery may dispose less volatile fluorine to their pond system, thus having fewer emissions from the ponds.

The most important emission source in the typical WPPA process is the ventilating air from the digester. The digester vent gases contain water vapor, particulate dust, and SiF_4. The removal of fluorine can be accomplished by the wet process or adsorption (solid reagent of adsorption system) of the ventilating air. The wet process has been used exclusively. However, the advantages and capabilities of the adsorption process should be evaluated [U.S. EPA, 1973].

The design of the scrubbing equipment is limited by the chemistry of the reactions between the gypsum pond water and the fluorine containing gases discharged from the WPPA plant reactor. The following reactions occur in the reactor:

$$CaF_2 + H_2SO_4 \rightleftarrows CaSO_4 + 2HF \tag{8.2}$$

or

$$2HF + SiF_4 \rightleftarrows H_2SiF_6 \tag{8.3}$$

Furthermore, hydrolysis of SiF_4 can occur at high concentrations, as follows:

$$3SiF_4 + 4H_2O \rightleftarrows Si(OH)_4 + 2H_2 + SiF_6 \quad (8.4)$$

As a result of this reaction, a gelatinous deposit of polymeric silica is formed, which tends to plug the scrubber packings [U.S. EPA, 1973].

The three most effective scrubbing techniques for removal of fluorine are: (1) counter-current, (2) co-current, and (3) cross-flow scrubbing. Counter-current scrubbers have the highest potential removal efficiency since they contact the gas leaving the scrubber with incoming clean liquor. However, co-current scrubbing is less prone to plugging, and requires a lower gas pressure drop to operate. The cross-flow scrubber is a compromise between the counter and co-current scrubbers. It combines the efficiency of the former with the mechanical advantages of the latter. The cross-flow scrubber has been used in combination with spray towers, venturi, or wet cyclones in phosphoric acid plants [U.S. EPA, 1973].

The efficiency of the scrubber is dependent on the temperature and composition of the scrubber medium. The gypsum pond water used in the scrubber contains 3000 to 10,000 ppm of fluorine [Nyers, et al. 1979]. Efficient removal of fluorine from the gas stream is reduced, due to the high partial pressure of hydrogen fluoride in the pond water. The mass transfer rate also decreases as temperature increase. Thus, fresh water should be used as the scrubbing medium in the last stage of scrubbing.

Oxides of sulfur are also emitted in the WPPA process. Their production ranges from 0.00777 to 0.058g/kg P_2O_5. However, the origin of SO_x formation is not clear in the WPPA process [Nyers, et al. 1979]. The cause of its emission must first be determined before measures regarding its reduction within the process are taken.

RECOMMENDED AREAS OF POLLUTION CONTROL RESEARCH: WET PROCESS PHOSPHORIC ACID PRODUCTION

After evaluating the process technology, the following areas are recommended for further research and development:

- Studies on purification of phosphate feed to reactor, to reduce impurities which cause by-product formation.

- Comparative studies on adsorption and absorption of fluorine, to determine the most efficient technique to alleviate fluorine emission.

- Research on the design and operating parameters of the scrubber, to reduce plugging and increase the rate of fluorine transfer from the vent gases to scrubbing medium.

NORMAL SUPERPHOSPHATE PRODUCTION

Normal superphosphate (NS) is produced by the reaction of phosphate rock with sulfuric acid. The phosphate rock and H_2SO_4 are mixed in a cone mixer (reactor vessel), the acidulate is then transferred to an enclosed area (den) to solidify. The solidified product is then stored for curing. A schematic diagram of this process is shown in Figure 8.2.

NS production can be accomplished by both continuous and batch processes. For the batch process, a pan mixer (instead of cone mixer for the continuous process) is used. However, both processes convert fluorapatite in the rock to soluble monocalcium phosphate. The major reaction in this process is:

$$[Ca_3(PO_4)_2]_3 + CaF_2 + 7H_2SO_4 + 3H_2O \rightleftarrows \\ 3[CaH_4(PO_4)_2 \cdot H_2O] + 7CaSO_4 + 2HF \quad (8.5)$$

The sources of pollution emissions for a NS plant are the mixers, den, and curing building. Between 1.5 kg and 9.0 kg of fluorides/metric ton of NS are released during production and curing [Nyers, et al. 1979]. The fluoride emissions occur in the form of fluoride vapor, evolving as hydrogen fluoride and silicon tetrafluoride. The fluoride emissions range fom 0.07 g F/kg P_2O_5 to 0.45 g F/kg P_2O_5. Individual emission rates from each unit are hard to determine since all gases are usually vented to the same stack.

As previously discussed, scrubbing is inhibited by the formation of a gelatinous mass,

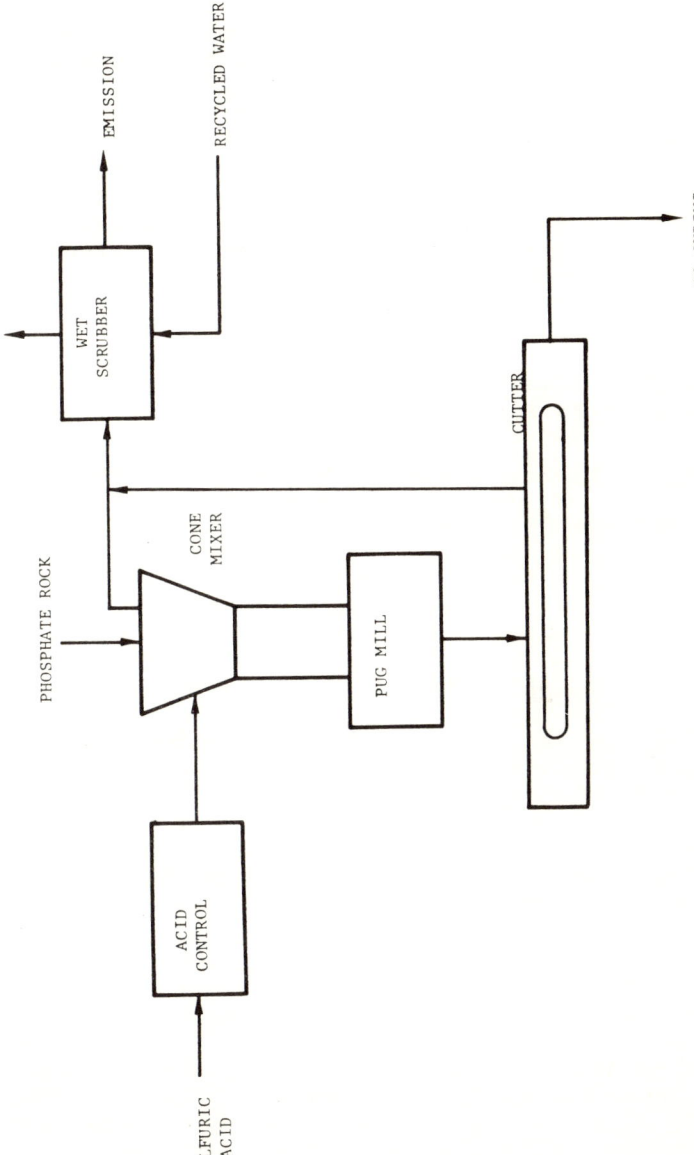

Figure 8.2. Flow Chart of Normal Superphosphate Production Process (Nyers, et al. 1979).

generated by the reaction of tetrafluoride and water. Thus, the use of conventional scrubbing, for pollution abatement, is greatly influenced by this reaction. The effectiveness of the fluoride reduction by a scrubber is determined by inlet fluorine concentration, outlet saturation temperature, composition and temperature of the scrubbing liquid, scrubber type and transfer units, and effectiveness of entrainment separation [Heller, et al. 1968].

The best strategy for pollution control is to use a cross-flow scrubber. Furthermore, fluorine removal is enhanced by use of fresh water in the last stage and increasing the number of stages in the scrubber [Nyers, et al. 1979].

RECOMMENDED AREAS FOR POLLUTION CONTROL RESEARCH: NORMAL SUPERPHOSPHATE PRODUCTION

After evaluating the process technology, the following area is recommended for further research and development:

- Studies on the scrubber design and optimal operating parameters to reduce fluorine emissions.

AMMONIUM PHOSPHATE PRODUCTION

The production of ammonium phosphate (AP), mostly diammonium phosphate (DAP), has increased rapidly in the past 3 decades. The total of ammonium phosphate produced in the U.S. in 1979 reached nearly 3.7 million tons P_2O_5, which is 63.4 percent of all phosphate fertilizers produced in the U.S. [Kirk-Othmer, 1980]. Approximately 99 percent of AP is used as fertilizer [Nyers, et al. 1979].

The discussion to follow includes the granulation of phosphoric acid with anhydrous ammoniation-granulation to produce granular fertilizer. Ammonium phosphate is produced by the reaction of phosphoric acid with anhydrous ammonia. In 1975, 84 percent of AP produced in the U.S. was of DAP grade (i.e percentage of available N-K-P is 16-0-48, [Nyers, et al. 1979].

Ammonium phosphate [$(NH_4)_2HPO_4$] is produced by the reaction of 1 mole of phosphoric acid with 2 moles of ammonia, forming a product with 21.1

percent nitrogen and 54 percent available phosphorous. Monoammonium phosphate (MAP) can also react with ammonia to yield DAP. The DAP production can be performed in either a pugmill or rotary drum ammoniator. Approximately 95 percent of ammoniation granulation plants in the U.S. use a rotary drum-mixer developed by TVA [Rawlings and Reznik, 1976].

The pollution sources arise from four operations in the ammonium production process: (1) reactor, (2) ammoniator, (3) dryer, and (4) cooling. Typical emissions include, ammonia, flouride, particulates, and a negligible amount of combustion gases. The following discussion is for the TVA process depicted in Figure 8.3.

The reactor or preneutralizer is a vessel into which 70 percent of ammonia and all of the phosphoric acid is introduced. The reactor operates at atmospheric pressure and 100-120 °C. The $NH_3:H_3PO_4$ ratio is maintained at 1.3 to 1.5:1 for maximum solubility of MAP, which is the main product of the reactor [Nyers, et al. 1979].

The emissions from the reactor include gaseous ammonia and fluorides. The emissions are caused by volatilization due to incomplete chemical reactions and excess free ammonia. Collective emissions for the reactor and ammoniator-granulator are as follows:

Fluorides (as F)	0.023g/kg P_2O_5
Particulate	0.076g/kg P_2O_5

Ammonia emission data is available for the AP production as a whole [Nyers, et al. 1979].

The reactor emissions can be controlled with a more efficient design of the reactor. In theory, the reactor can be designed without any ventilation, but in practice they are operated at a 57-76 cfm ventilation rate [U.S. EPA, 1973].

The reactor design can be modified to reduce the ventilation rate and the atmospheric emission. Furthermore, excess ammonia should be avoided in the reactor vessel. The rate of reaction could be increased catalytically by changing the process conditions, and/or the residence time should be increased in order to achieve complete reaction.

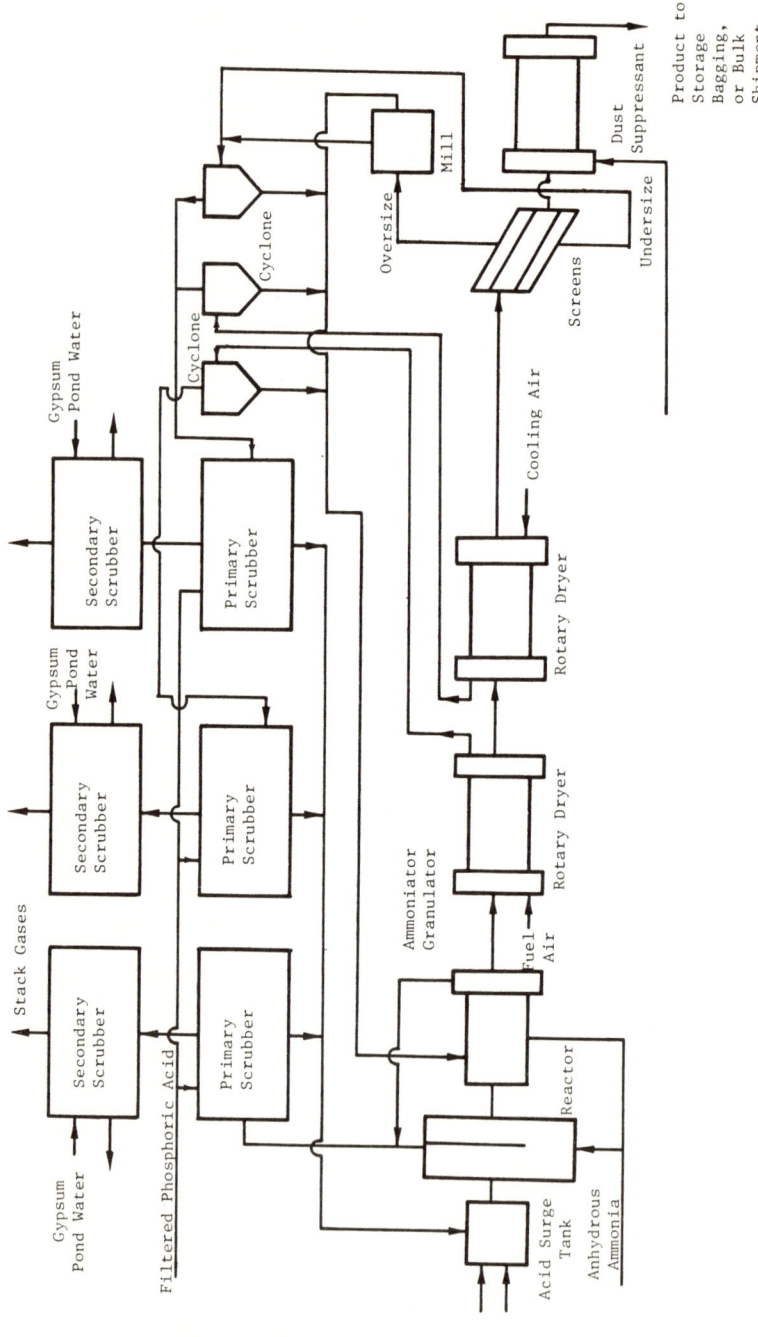

Figure 8.3. Flow Diagram for TVA Ammonium Phosphate Process. (Nyers, et al. 1979).

The granulation by agglomeration and by coating particles with slurry from the reactor takes place in the rotating drum. Ammonia is supplied by a sparging action underneath the bed, to bring the $NH_3:H_3PO_4$ ratio to 1.8 to 2.0:1.0 [U.S. EPA, 1973]. The granulation can, theoretically, be performed with no ventilation. The ventilation is mainly a function of design of the granulator. Thus, design considerations play an important role in the emissions of ammonia.

The emissions from the reactor and granulator are scrubbed for pollution reductions. Any one of the various scrubbing techniques can be applied to these off gases. However, packed scrubbers should be avoided due to the formation of gelatinous silicon or DAP, which tend to plug the scrubber. The best scrubbing technique of primary scrubbing is a venturi cyclone scrubber, which has good capabilities for absorption of ammonia and capture of particulate matter.

Moist granules of DAP are dried to 2 percent moisture in a counter-current gas or oil fired dryer unit. The controlled emissions from the dryer and cooler are 0.015g/kg P_2O_5 fluoride and 0.75g/kg P_2O_5 particulate emissions [Nyers, et al. 1979].

The fluoride emissions are due to the dissociation of the fertilizer product, and particulate emissions are caused by entrainment of DAP and MAP dusts in the ventilation air streams. The particulate emissions may contain both ammonium fluoride and ammonium phosilicates [U.S. EPA, 1973]. Typically, the primary scrubber is designed to capture the particulates. Thus, the design of the primary scrubber must account for the particulate as well as removal of off gases from the ventilation streams. Furthermore, the scrubbing medium, in the primary scrubber, should be acidic enough to absorb the ammonium discharge. For this purpose 30 percent phosphoric acid solution is used for the primary scrubbing medium, to recover ammonia [U.S. EPA, 1973].

When the use of only one scrubber is not sufficient to remove particulates and gases, combination of scrubbers must be considered.

Table 8-1. Pollutants and Suggested Strategy for Selected Phosphate Rock Fertilizer

Industry	Process	Pollutants	Sources in Process	Nature of Pollutants	Pollutant Control Strategy
Wet process phosphoric acid	Phosphate Rock Processing	Flourides Particulates, SO_x, & Gypsum.	Reactor vessel, filteration, vaporation, & gypsum pond	Inorganic gases, and solids	Crossflow scrubbing with fresh water in the last stage, and/or use combination of scrubbers.
Normal Super-Phosphate Production	Phosphate Rock Processing	Flourides, Particulates	Mixers, den, and curing.	Inorganic gases, and solids	Scrubbing the offgases, cross flow scrubbing, and/or use combination scrubbing
Ammonium Phosphate Production	Diammonium Phosphate Production	Ammonia, Flourides, and Particulates	Reactor, ammoniator, dryer, and cooler.	Inorganic gases, and solids	Use combination primary and secondary scrubbing improve the design on the reactor and ammoniator to reduce ventilation rate.

Table 8.1 contains a summary of pollution problems associated wiht the phosphate rock industry and the suggested strategy.

RECOMMENDED AREAS FOR POLLUTION CONTROL RESEARCH: AMMONIUM PHOSPHATE PRODUCTION

After evaluating the process technology the following areas are recommended for further research and development:

- Research and kinetics of reaction of ammonia with phosphoric acid to enhance this reaction, either catalytically or by increasing the residence time in the reactor vessel, which would reduce the emission.

- Studies on design of the reactor vessel and granulators to minimize the ventilation rate, which would lead to smaller volumes of gaseous emission.

- Research on the design and selection of optimal operating parameters for the scrubbing unit, to reduce the ammonia, fluoride, and particulate emissions.

REFERENCES

Chemistry and Technology of Fertilizer. V. Sauchelli, ed. Rienhold Publishing Corp., New York, New York 1960.

Corbridge, D. Phosphorous, Elsevior Scientific Publishing Company, New York, 1978.

Heller, A.N., Cuffi, S.T. and Goodwin, D.R. Inorganic Chemical Industry, Air Pollution, Vol. III: Source of Air Pollution and Their Control, A.C. Stern, ed., Academic Press, New York, New York, 1968.

Lutz, W.A. and Pratt, C.J. Principles of Design and Operation in: Phosphoric Acid, No. 1, A.V. Slack, ed. Marcel Dekker, Inc., New York, New York, 1968.

Nyers, J., et al Source Assessment: Phosphate Fertilizer Industry. EPA-600/2-79-019C, May 1979.

Olson, R., et al. Fertilizer Technology & Use.
Soil Science Society of America, Inc.,
Wisconsin, 1971.

Rawlings, G.D. and Reznik, R.B. Source Assessment:
Fertilizer Mixing Plants. EPA-600/2-76-032C.
March 1976.

Slack, A.V. Chemistry and Technology of
Fertilizers. John Wiley and Sons, Inc., New
York, New York 1967.

U.S. EPA. Air Pollution Control Technology and
Costs in Seven Selected Areas. Industrial Gas
Cleaning Inst., Inc., Stanford, CT, EPA-450/3-
73-010, March, 1973.

U.S. EPA. Background Information for Standards of
Performance: Phosphate Fertilizer Industry,
Environmental Protection Agency, Research
Triangle Park, N.C. Office of Air Quality
Planning and Standards. EPA-450/2-74-019A,
Oct. 1974.

SUPPLEMENTAL REFERENCES

Gartrell, F.E. and Barber, J.C. Pollution Control
Interrelationships, Chemical Engineering
Progress, 62(10):44-77, 1966.

Huffstufler, K.K. Pollution Problems in Phosphoric
Acid Production in: Phosphoric Acid, Vol. I,
A.V. Slack, ed., Marcel Dekker, Inc., New
York, New York, 1968.

Illarionov, N.W., et al. Zh. Prikl-Khim. Vol 36,
1963.

Kirk-Othmer Encyclopedia of Chemical Technology,
Third Edition, Volume 101, John Wiley and
Sons, New York, 1980.

Lehr, J.R. Purification of Wet Process Acid In:
Phosphoric Acid, Volume I, A.V. Slack, ed.
Marcel Dekker, Inc., New York, 1968.

Muehberg, P.E., Reding, J.T. and Shepherd, B.P.
Draft Report: The Phosphate Rock and Basic
Fertilizer Materials Industry. Contract 68-
02-1324, Task 8, EPA, Research Triangle Park,
North Carolina, May 1976.

Shreve, R.N. *Chemical Process Industries*, Third Edition. McGraw-Hill Book Company, New York, 1967.

U.S. EPA. *Atmospheric Emissions from Wet Process Phosphoric Acid Manufacture*, U.S. Department of Health, Education and Welfare, NAPLA. No. AP-57, April 1970.

U.S. EPA. *Final Guideline Document: Control of Fluoride Emissions form Existing Phosphate Fertilizer Plants*. EPA 450/2-77-0051 March 1977.

INDEX

acid pickling
  combination 89
  hydrochloric 89
  sulfuric 89
aluminium 117
ammonia 54
ammonium phosphate 140

bauxite 117

carbon cathode 124
catalyst 36
cementation 13
coke 71
combustion 107

dezincing 83
digester system
  batch 101
  continuous 101
dust
  dry 83
  wet 83

electroplating 25
electrorefining 7
explosives
  nitrocellulose 51
  trinitrotoluene 51

fertilizers 133
finishing 61
flotation 5
  comminution 5
  froth 5

gas
  acid 36
  purification 41
  upgrading 41
gasification 39

hydrometallurgical 11

iron making 79

leaching 12
  agitated 12
  dump 12
  in situ 12
  vat 12
lime kiln 110
liquefaction 31

mercaptants 101
methyl 101

nitration 58
nitric acid 51
  concentration 55
  production 51
nitrogen oxide 37
nonferrous metals 3
normal super-
  phosphate 140

phosphoric acid   133
pink water   60
pollutants
  dissolved metal salts   3
  particulates   3
  slag   3
  SO_2 3
production
  copper   3
  uranium   3
pulp wood   99
  cellulose   99
  lignin   99
pulping   99
purification   60

recovery furnace system   105
red water   60

slags   84
smeltings   5
solvent extraction   13
solvent refined coal   32
steel   71
steel making   85
  basic oxygen furnace   86
  open hearth furnace   87
  electric arc furnace   87

washers
  pressure   103
  vacuum   103

yellow water   60